SpringerBriefs in Business

More information about this series at http://www.springer.com/series/8860

Elias G. Carayannis • David F. J. Campbell

Smart Quintuple Helix Innovation Systems

How Social Ecology and Environmental Protection are Driving Innovation, Sustainable Development and Economic Growth

Springer

Elias G. Carayannis
School of Business
George Washington University
Washington, DC, USA

David F. J. Campbell
Center for Educational Management
and Higher Education Development
Department for Continuing Education
Research and Educational Management
Danube University Krems
Krems, Austria

Unit for Quality Enhancement/UQE
University for Applied Arts Vienna
Vienna, Austria

ISSN 2191-5482 ISSN 2191-5490 (electronic)
SpringerBriefs in Business
ISBN 978-3-030-01516-9 ISBN 978-3-030-01517-6 (eBook)
https://doi.org/10.1007/978-3-030-01517-6

Library of Congress Control Number: 2018956576

This Springer imprint is published by the registered company Springer Nature Switzerland AG
The registered company address is: Gewerbestrasse 11, 6330 Cham, Switzerland

Contents

About the Authors

Elias G. Carayannis is full professor of science, technology, innovation, and entrepreneurship; co-founder and co-director of the Global and Entrepreneurial Finance Research Institute (GEFRI); and director of research on science, technology, innovation, and entrepreneurship, European Union Research Center (EURC), at the School of Business of the George Washington University in Washington, D.C. His teaching and research activities focus on the areas of strategic government-university-industry R&D partnerships, technology road-mapping, technology transfer and commercialization, international science and technology policy, technological entrepreneurship, and regional economic development.

David F. J. Campbell is associate professor (Privatdozent) in the Department of Political Science at the University of Vienna; a faculty member at the Center for Educational Management and Higher Education Development, Department for Continuing Education Research and Educational Management, at Danube University Krems; a quality expert at the University of Applied Arts in Vienna; and a researcher (Senior Scientist) with the Alpen-Adria-University of Klagenfurt.

Abstract

The starting point for analysis is the key relationship between ecology and innovation and economic development, as it is being postulated and suggested by the principles of the Quintuple Helix innovation systems: "Within the framework of the Quintuple Helix innovation model, the natural environments of society and the economy also should be seen as drivers for knowledge production and innovation, therefore defining opportunities for the knowledge economy. ... The Quintuple Helix supports here the formation of a win-win situation between ecology, knowledge and innovation, creating synergies between economy, society, and democracy. Global warming represents an area of ecological concern, to which the Quintuple Helix innovation model can be applied with greater potential" (Carayannis, Barth and Campbell, 2012, p. 1, http://link.springer.com/article/10.1186/2192-5372-1-2). Recent empirical evidence suggests further that there is (by tendency) a decoupling of economic growth from a further increase of energy sector emissions: "Private-sector incentives help drive decoupling of emissions and economic growth. ... The importance of this trend cannot be understated. This 'decoupling' of energy sector emissions and economic growth should put to rest the argument that combatting climate change requires accepting lower growth or a lower standard of living" (Obama, 2017, p. 1, http://science.sciencemag.org/content/early/2017/01/06/science.aam6284.full). Finally, toward the end of analysis, key features for a program of Smart Quintuple Helix Innovation Systems are being presented for discussion.

Chapter 1
Introduction

"Global warming" represents an ecological (also socio-ecological) issue of importance and concern. Currently it can be stated that concentrations of CO2 emissions in the atmosphere are reaching historical all-time highs, which is causing severe ecological and environmental problems, for example, global warming (World Meteorological Organization, 2017). Due to the escalation of global warming, it is time for humanity to think and act responsibly and determine sustainable solutions. Global warming, in addition to climate change, has caused the world to undertake new responsibilities (see IPCC, 2007a, 2007b), which not only include further climate change but in the long term also hold humanity accountable in the prevention of new political and/or social conflicts, war on resources, new environmental catastrophes, as well as serious crises in the market economies (see UNDP, 2007; UNEP, 2008). The special challenge of global warming can be tackled by "sustainable development."[1] Sustainable development concerns us all and takes place on the local as well as global level. Hence, sustainable development has to be understood in the context of "gloCal knowledge economy and society" (see Carayannis & Campbell, 2011; Carayannis & Alexander, 2006; Carayannis & Von Zedtwitz, 2005). Therefore, we must perceive global warming not as a challenge but rather as an opportunity to live innovatively and effectively in union with nature for a better tomorrow.

To a large extent, humanity itself has caused the climate change; therefore something must be done (see Friedman, 2008; IPCC, 2007b; Le Monde diplomatique, 2009, pp. 72–73). However, there are hardly any comprehensive models or concepts to answer the "WHY" that truly show "HOW" we can act and learn accordingly or provide any demonstrative methods, suggestions, and examples "HOW" we can improve our actions in the present. Our analysis presented here suggests

[1] The definition of the Brundtland Commission states that sustainable development "meets the needs of the present without compromising the ability of future generations to meet their own needs" (United Nations 1987a, 1987b).

E. G. Carayannis, D. F. J. Campbell, *Smart Quintuple Helix Innovation Systems*, SpringerBriefs in Business, https://doi.org/10.1007/978-3-030-01517-6_1

understanding the "WHY" and consequently offers a "model of innovation," which demonstrates a feasible, step-by-step method to tackle the "HOW."

In the current academic debate, it is undisputed that a solution or a suitable answer regarding the challenge of global warming can only be found through utilizing the asset of human knowledge (see Bhaskar, 2010, p. 1; Carayannis & Campbell, 2010, p. 42). The key to success, as is being determined by our propositions, lies in using the available and newly created "knowledge" in correspondence with the *Quintuple Helix model* (Carayannis & Campbell, 2010, p. 62). The *Quintuple Helix* is a model of innovation that can tackle existing challenges of global warming through the application of knowledge and know-how as it focuses on the social (societal) exchange and transfer of knowledge inside subsystems of a specific state, nation-state (see Barth, 2011a, pp. 5–7; see furthermore Barth, 2011b, 2011c). The "nonlinear" innovation model of the *Quintuple Helix*, which combines knowledge, know-how, and the natural environment system together into one "interdisciplinary" and "transdisciplinary" framework, can provide a step-by-step model to comprehend the quality-based management of effective development, to recover a balance with nature, and to allow future generations a life of plurality and diversity on earth (see Barth, 2011a, p. 2; Carayannis & Campbell, 2010, p. 42). To sum up, our thesis is as follows: the *Quintuple Helix* represents a suitable model in theory and practice offered to society, to understand the link between knowledge and innovation, in order to promote lasting development.

> This analysis is being guided by the following key research question: *How do knowledge, innovation, and the environment (natural environment) relate to each other?* The second research question (to be addressed in the conclusion) is: *What are key features of Smart Quintuple Helix Innovation Systems?*

Advanced or advancing knowledge and innovation systems (across a multi-level architecture of sub-national, national, and transnational levels) could be characterized by a pluralism of knowledge and innovation modes. In fact, a certain co-evolution or congruence between advanced knowledge (innovation) systems and advanced (high-quality) democracy may be stated, postulating that advanced knowledge and innovation take over some of the structural elements of a democracy, such as pluralism and diversity.

Referring to the research question as conceptual point of departure, our final objective is to design and to propose for discussion an interdisciplinary and transdisciplinary framework of analysis for sustainable development and social ecology that exactly ties together knowledge, innovation, and the environment. This model we will call the Quintuple Helix, a five-helix model that embeds the Triple Helix and the Quadruple Helix. Triple Helix focuses on knowledge production and use in context of "university-industry-government relations" (Etzkowitz & Leydesdorff, 2000). Quadruple Helix extends the Triple Helix by adding the helix of a "media-based and culture-based public" (Carayannis & Campbell, 2009). The Quintuple Helix contextualizes the Triple Helix and Quadruple Helix by further adding on the helix of the "environment" ("natural environments"). The Quintuple Helix thus offers an analytical frame or framework where knowledge and innovation, on the one hand, are being connected with the environment, on the other. By this the Quintuple Helix addresses and incorporates features of "social ecology." Furthermore, the Quintuple Helix also can be seen as a framework for interdisciplinary analysis and transdisciplinary problem-solving in relation to sustainable development, because a comprehensive understanding of the Quintuple Helix clearly implies that knowledge production and use as well as innovation must be set in context or must be contextualized by the natural environment of society.

The concept of Triple Helix innovation systems was introduced by Etzkowitz and Leydesdorff (e.g., see Etzkowitz & Leydesdorff, 2000). The metaphor of *Helix* or *Helices (Helices, spirals)* refers here to interwoven and cross-connected and cross-interconnected sectors. Triple Helix is possible within a democracy. However, Triple Helix is also possible without a democracy. The Triple Helix focuses on the knowledge economy, which may be approached by a democratic or a nondemocratic political framework. Nondemocratic (authoritarian) political regimes may be tempted to implement varieties of Triple Helix designs. Per definition, to already begin with a conceptual starting point, it is impossible for a nondemocratic (authoritarian) political regime trying to implement a Quadruple Helix (Carayannis & Campbell, 2012). *There is no Quadruple Helix without democracy* (Campbell & Carayannis, 2013a, 2015; Campbell, Carayannis, & Rehman, 2015). In addition, evidence suggests that the ecological sensitivity of the Quintuple Helix (Carayannis, Barth, & Campbell, 2012) can be more easily or realistically implemented and promoted within a democratic context of knowledge production and innovation. For the Quadruple Helix the "democracy matters": *this is in line with a view of a "Neo-Renaissance" where democracy encourages development in action for smart, sustainable, and inclusive growth, by this advocating sustainable development.* This should allow for "happy accidents" (Carayannis, Campbell, & Rehman, 2016). For discourses on knowledge and innovation, a *democracy* versus *technocracy* issue can be postulated, where technocratic (and bureaucratic) approaches to innovation in nondemocratic regimes are being questioned and challenged by knowledge production and innovation in democracies. Also for the developing countries and emerging markets, this has implications and ramifications, where there should be expectations that developments in knowledge and innovation are paralleled by progress in democratization (of course, this may not be always the case in empirical terms or empirically). Democracy acts as one of the levers that "happy accidents" in knowledge production and innovation are being transformed and translated into opportunities and benefits for society and to the people. Can there also be a "democratic capitalism," and which attempts of realization can there be approached or tried out (Carayannis & Kaloudis, 2010)?

The structure of the book is as follows. In Chap. 2, several key terms are being defined, referring to knowledge, innovation, and democracy. Chapter 3 discusses these key terms in greater detail, and in Chap. 4, the Quintuple Helix is being explained and developed further. Chapter 5 finally reruns the earlier analysis at a more advanced level (*in the mode of a "Helix"*), by moving and by progressing *Quintuple Helix* to the *Quintuple Helix Innovation Systems*. In the conclusion (Chap. 6), the analysis refers again to the key questions that were raised in the introduction (This Chapter).

Chapter 2
Definition of Key Terms: Knowledge, Knowledge Production, Innovation, Democracy, and Governance

The Wikipedia definition of knowledge, also cross-referencing to the Oxford English dictionary, lists as a crucial element of knowledge "the theoretical or practical understanding of a subject." The Wikipedia definition furthermore associates knowledge to "expertise, and skills" that a person may have gained either by experience or through education.[1] Currently, there exists a general belief (indicated by numerous publications) that knowledge becomes increasingly important for society, economy, and also democracy. Advancements and a sustainable development of society and the economy appear unlikely without leveraging and enhancing knowledge. This adds plausibility for using concepts such as the *knowledge-based society*, the *knowledge-based economy*, and the *knowledge-based democracy* (Carayannis & Campbell, 2009, p. 224). Perhaps there is even a shift not only to speak of the knowledge-based society and economy but of a *knowledge society* and a *knowledge economy* per se that is being endogenously driven by knowledge. The concept of a *knowledge democracy* consequently complements such propositions.

One could set up two conceptual axes, trying to model knowledge and different types of knowledge in greater detail (Campbell, 2009). One axis may polarize "codified" (explicit) with "tacit" knowledge (see, e.g., Gibbons et al., 1994, p. 167–168). Tacit knowledge represents an experience-based knowledge, whereas codified knowledge is written down in the one or other form. The other axis could polarize knowledge that is less dependent or more dependent on the context: a possible conceptual wording would juxtapose (compare) knowledge that is "independent of users and/or appliers" with a knowledge that is "dependent of users and/or appliers." Here differing degrees of *contextualization of knowledge* become manifest and evident. The closer a knowledge places to "codified" and "user-independent," the more this knowledge is "information." Contrarily, the closer a knowledge places to the poles of tacit and user-dependent, the more various types of "competencies" are being expressed. Competencies again stretch from professional or expert knowledge (know-how) to social competencies (soft skills, intercultural competencies)

[1] See http://en.wikipedia.org/wiki/Knowledge

E. G. Carayannis, D. F. J. Campbell, *Smart Quintuple Helix Innovation Systems*, SpringerBriefs in Business, https://doi.org/10.1007/978-3-030-01517-6_2

and competences of the personality.[2] Higher education teaching, currently, stresses the notion of "desired learning outcomes" that become visible as competencies acquired by students. Schafer (2008, p. 276) discusses prospects of a new "era of balanced leadership" in context of political leadership: this clearly would require mature social competences on the part of politicians.

Complementary to the above-depicted modeling of knowledge based on the two axes of codified/tacit and user-independent/user-dependent, an alternative modeling could focus more on aggregated features of knowledge, emphasizing systemic aspects and embedding knowledge in a larger societal context. Here several axes (or dimensions) may be discussed for a broader systemic approach (see also Carayannis & Campbell, 2009, p. 214–215):

1. Research, R&D (research and experimental development): conventionally, research is being distinguished in basic research, applied research and experimental development (OECD, 1994, p. 29; OECD, 2002, p. 30).[3]
2. Education: education can refer to primary education, secondary education, and tertiary education, where tertiary education is the education being offered by universities or the higher education system (containing all HEIs, the higher education institutions) in more general.
3. Innovation.
4. Different spatial axes, which represent geographic, geographic-spatial, or spatial-political concepts, distinguishing between the sub-national (local), national, and transnational (supranational, global) levels.
5. Perhaps also other nonspatial axes would be possible, for example, "creativity" and attempts of displaying and measuring creativity.

Focusing on research (R&D), the so-called linear model of innovation was prevailing for a long time. This linear model leverages on the fact that the universities (the HEIs) concentrate on basic research (often or mostly publicly funded), while firms concentrate on experimental development (often or mostly privately financed) (for the USA, see National Science Board, 2008, Volume 1, Chapter 4, p. 14–15). Applied research often is being seen to position itself "between" basic research and experimental development. This is carried by the underlying understanding that ideas, products or services start as a basic research in context of universities, and gradually diffuse time-lagged into society and the economy. Firms selectively pick up some basic research results and convert these through applied research and experimental development into commercially profitable products or services for the market. Challenges obviously are how to design systematically interfaces and linkages between publicly funded basic university research and the privately funded firm-based commercialization of research for profitable business activities. Potential risks (or market failures) could be a private underinvestment of research or of basic research (Tassey, 2001, p. 42, 61–64). Kline and Rosenberg (1986) and Miyata (2003, p. 715) describe the linear model as a sequence of the following concepts:

[2] In that understanding, emotional competence may crosscut social competences and personality.

[3] The OECD is the "Organisation for Economic Co-operation and Development.".

basic research, applied research, development, production, and marketing. This "simple linear model" Narin, Hamilton, and Olivastro (1997, p. 318) summarize as "The notion that technology springs from a scientific base was originally embedded in the 'linear model' of innovation: from basic research through applied research continuing into technology and resultant economic benefit." Lundvall (1992, p. 13) paraphrases this also as a "linear model of technical change." Interestingly, this linear model of innovation often is being closely associated with Vannevar Bush and his pivotal report *Science The Endless Frontier* (Bush, 1945, see the chapter on "The Importance of Basic Research"). Narin et al. (1997, 317–318) claim this implicitly, and in substance this association of the linear innovation model to Bush appears correct. At the same time, however, it should be mentioned that Bush himself, in his famous text, not even mentions the word "innovation" (as can easily be verified by an electronic word search command). Currently, the linear innovation models are being challenged by nonlinear innovation models that stress the importance of a simultaneous coupling of basic (university) research with the commercial R&D applications of firms in the business sectors. Kline and Rosenberg (1986) propose to introduce here a so-called chain-linked model (see also Miyata, 2003, p. 716). The underlying concept is to cross-link mutually and directly basic university research and the applied R&D commercialization in firms but also to foster basic business research and applied research in universities (Carayannis & Campbell, 2009, p. 209–211). In metaphoric terms, the *first-then* ("zuerst-dann") relationships in the linear model are being extended by *simultaneously-simultaneously* ("gleichzeitig-gleichzeitig") relations and network configurations in nonlinear knowledge arrangements. Originally *sequenced* processes are being *parallelized* (Carayannis & Campbell, 2009, p. 217; Campbell & Güttel, 2005, p. 167–168).

In a systemic (social, societal) understanding, knowledge creation and knowledge production often are associated more closely to research, basic research, and the sciences; thus a function of universities (HEIs), embedded in a national or multi-level innovation system, is to focus exactly on knowledge creation and knowledge production.[4] Of course, also other organizations, such as firms, can focus and specialize on knowledge production. Knowledge creation and production are being complemented by the concepts of knowledge application, knowledge diffusion, and knowledge use. This could imply to think of *two sides* of knowledge: knowledge creation and production on the one hand, and knowledge application and use on the other (see Fig. 2.1). Knowledge application and knowledge use already overlap substantially with the concept of innovation that could be defined as: *innovation leverages knowledge for knowledge application, diffusion, and use and thus translates knowledge into application*. This definition of innovation has references to knowledge and leaves the question open (and unresolved), whether there could be an innovation without knowledge (Carayannis & Campbell, 2009, p. 213–214). There

[4] In the context of this analysis, we use "knowledge creation" and "knowledge production" as interchangeable terms. A possible distinction may emphasize that knowledge creation is more fundamental and basic (more overlapping with basic research) than the knowledge production.

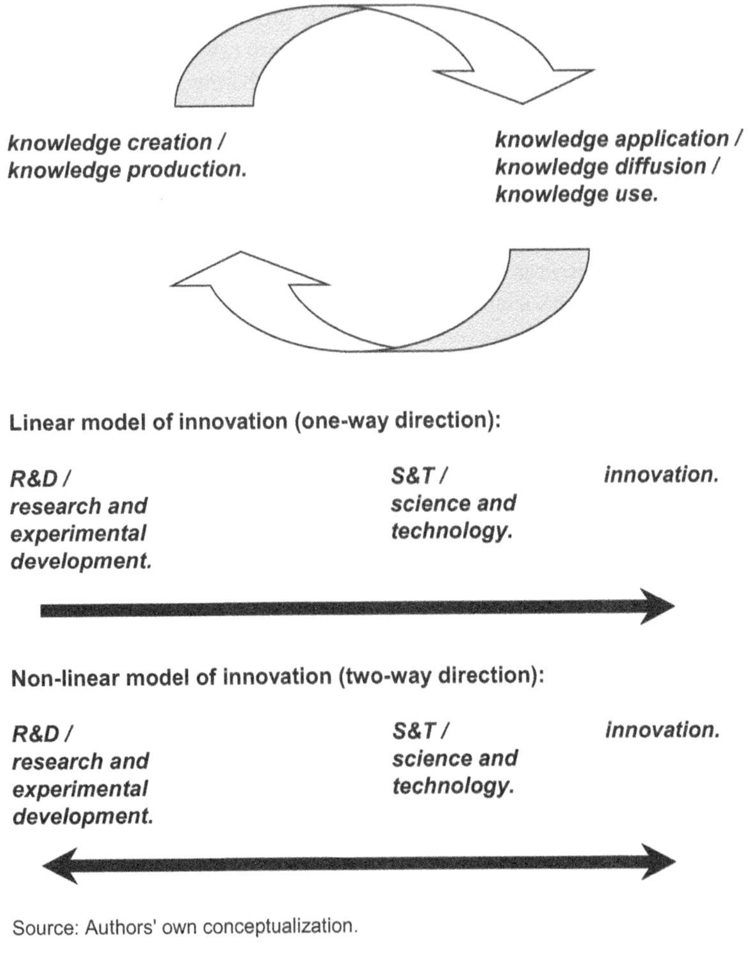

Fig. 2.1 Conceptualization of (a possible) relationship of knowledge and innovation. Source: Authors' own conceptualization

exists basic research,[5] "pure research," or "pure science"[6] that is not interested in issues of application and innovation. Whether an innovation, for example, some forms of management innovations in business that are not R&D or technology-based,

[5] We want to quote, how the OECD (1994, p. 29) defines basic research: "Basic research is experimental or theoretical work undertaken primarily to acquire new knowledge of the underlying foundation of phenomena and observable facts, without any particular application or use in view." This definition the OECD repeats unchanged in 2002 (OECD, 2002, p. 30).

[6] Etzkowitz and Leydesdorff (2000, p. 116) use the term of "pure science" for describing the post-1945 university system in the USA, which largely behaved according to the principles that were formulated and postulated by Vannevar Bush (1945). Etzkowitz and Leydesdorff (2000, p. 116) speak in this context also of an "ideology of pure research."

can qualify as an innovation without linkages to knowledge and could be debated. But there can be innovations that are not connected to basic research (for an overview on innovation, see Shavinina, 2003). S&T, science and technology, also spans to both poles of knowledge: science locates more closely to knowledge creation and production, while technology associates closer to knowledge application, use, and innovation. Technology may be interpreted as a type of innovation (often with a technological hardware component), interested in converting science (basic research) into commercial application and use.

The concept of the "national system of innovation" (or national innovation system) was developed by the two scholars Bengt-Åke Lundvall (1992) and Nelson (1993). This approach contextualizes innovation in the context of societies at the level of nation states. Lundvall (1992, p. 2) offers the following key definition: "It follows that a system of innovation is constituted by elements and relationships which interact in the production, diffusion, and use of new, and economically useful, knowledge and that a national system encompasses elements and relationships, either located within or rooted inside the borders of a nation state." For Lundvall (1992, p. 1) knowledge constitutes the "most fundamental resource" and learning the "most important process" in a modern economy. In that line of argument, we might postulate the following relationship: (1) innovation leverages or translates knowledge into application and use; (2) applied or used knowledge always or often or at least potentially may also be used economically for economic purposes, for generating financial revenues and profit; and (3) thus innovation also converts (potentially) knowledge creation and production into economic activities.

Lundvall (1992, p. 3–4) acknowledges that processes of globalization and regionalization weaken the national systems: "international specialization was often reflected in a regional specialization within the countries."[7] Despite the recognition of such sub-national and transnational innovation processes, Lundvall emphasizes that national patterns still exist and still play a key role, providing continued plausibility for the concept of the national innovation system: "… we believe that national systems still play an important role in supporting and directing processes of innovation and learning." The modern nation states acted as "engines of growth" (Lundvall, 1992).[8] Stefan Kuhlmann (2001, p. 972) also stresses the dominance of nation-state structures for the current political systems: "political systems are still nationally based, but are, in Europe, spreading increasingly both to the transnational and to the regional level." Lundvall (1992, p. 5) diagnoses that the concepts of the national innovation systems "already entered the everyday vocabulary of policymakers." This supports opportunities of and for a cross-country learning. Interestingly,

---

[7] In a different book section, Lundvall (1992, p. 15) says: "As pointed out, we do not assume the process of innovation to be exclusively localized inside national borders. On the contrary, we recognize that the process of innovation has increasingly become multinational and transnational reflecting, for example, R&D cooperation between big firms based in different nations."

[8] Nelson (1990) describes or paraphrases *capitalism as an engine of growth*. As Nelson (1990, p. 193) states at the beginning of his article: "Economists, from Marx, to Schumpeter, have touted capitalism as an engine of technical progress. But what kind of an engine is it? How does it work? What are the strengths and weaknesses?"

depending on the level of aggregation or the level of analysis, as Lundvall (1992, p. 7) says, innovation systems might differ in their ambitions and goals: the national level emphasizes "international competitiveness of the national economy"; at the level of international organizations, the efforts concentrate on "strengthening economic growth" and "avoiding international conflicts"; at the global level, ambitions focus on the "long-term survival of the global economy" that depend on "ecological sustainability" and a "reduction of the extreme social inequality." In the conceptual framing of Lundvall, moving bottom-up, the forces of learning gain in importance.

In further reflection of the concept of the *national innovation system,* now the concept of *multi-level systems of innovation* has entered the discourse.[9] Multi-level systems of innovation may be based on a geographic, spatial, geographic-political, or spatial-political understanding, juxtaposing, for example, sub-national, national, and transnational levels in one framework. Kuhlmann (2001, p. 970–971, 973) speaks of "multi-level, multi-actor systems" and of "multi-level innovation policy." Robert Kaiser and Heiko Prange (2004, p. 395, 405–406) use the terms of a "multi-level governance system" and discuss perspectives "from national to multi-level innovation systems." In addition to such "spatial axes," a multi-level system of innovation also could be based on "nonspatial axes" or, to be more precise, on nonspatial axes of knowledge aggregation (Campbell, 2006, p. 70; Carayannis & Campbell, 2009, p. 214–216). For example, innovation may be regarded as the highest form of knowledge aggregation of research (of the axis of research). Conventionally understood, technology is broader than research, and innovation again is broader than technology (Campbell & Güttel, 2005, p. 154; Carayannis & Campbell, 2006, p. 14–15). In that line of argument, innovation may also qualify as the broadest aggregation of knowledge of education (the axis of education). Rephrasing the above said, a multi-level innovation system could be based on several spatial and nonspatial axes that display different levels of (spatial and nonspatial) knowledge aggregation.

The one proposition for a definition for innovation is that of *converting knowledge creation and production to knowledge application, diffusion, and use.* From that logic it follows that, in principle, everything may qualify as belonging to a national (or multi-level) system of innovation that supports such processes and structures of knowledge application. How narrowly or how broadly (national) innovation systems are being defined, therefore, will differ and is interdependent with a concrete historical context. Depending on whether we believe or not believe in that an institution or structure should be associated to knowledge and innovation processes, this institution or structure would play a function for innovation and thus would be a part (or not) of the (national) innovation system. In a society, where knowledge is being associated primarily with knowledge creation and production in context of universities and higher education systems, and only few structural

[9] We can speculate, to which extent research about the European Union (EU) and concepts such as a *multi-level governance* of the EU (Bomberg & Stubb, 2003, p. 9; Hooghe & Marks, 2001) helped inducing and creating the concept of *multi-level systems of innovation* (Carayannis & Campbell, 2006, p. 11).

linkages to society and the economy, the "extension" of a national innovation system is more limited. In a society and economy, emphasizing knowledge application, diffusion, and use, the national innovation system obviously "broadens" and becomes increasingly powerful. Even culture (at least partially) could belong to the innovation system. Kuhlmann (2001, p. 954, 958, 967), for example, speaks of "innovation cultures," thus going clearly beyond a primarily institutional approach. In the words of Lundvall (1992, p. 12–23), "In different historical periods different parts of the economic system, or different inter-faces between subsystems, may play a more or less important role in the process of innovation."

The more knowledge application and knowledge use represent issues of interest, in a practical sense but also theoretically and conceptually, the more encompassing the national or multi-level innovation systems behave. The ultimate ratio would be that of a society or an economy that convert to a full and real knowledge society and knowledge economy, where almost everyone acts also as a knowledge worker, and with an innovation system stretching far out and into the peripheries of society. Kuhlmann (2001, p. 972) accentuates aspects of a "co-evolution" between the "political systems" and "innovation systems." The innovation system also could be understood as a system that cross-cuts into other systems of society, such as the political system, the research (R&D) system, the education system, and the economic system. The political system may try to influence the economic system directly with economic policy-making. Alternatively, the political system could be inclined to impose effects on the economic system via innovation policy. Innovation policy, then, would be an economic policy that cross-references to knowledge and that leverages knowledge: "Through innovation policy, however, which recognizes more specifically the conditions and ramifications of knowledge, the political system also projects an indirect and mediated, knowledge-tailored influence on the economic system. *This understanding underscores the interpretation and valuation of the innovation system as an interface between politics and the economy*" (Carayannis & Campbell, 2006, p. 18, 16–19).

What results, so the proposition, are forms of an "indirect coupling" between politics and the economy? One could even hypothesize that the more advanced a society and economy progress, we should expect at least a partial conversion or transformation from economic policy to innovation policy. Innovation policy represents a further developed form of economic policy in context of the knowledge society and knowledge economy. "For an advanced, knowledge-based democracy, knowledge and innovation policies qualify as a superior next-stage development of 'old' economic policies, and the degree of conversion from economic to knowledge and innovation policies may serve as a 'maturity test' for governance and policy-making" (Carayannis & Campbell, 2007, p. 87–88). However, we also want to stress that there are manifold opportunities for innovation policy that are not necessarily associated with economic activities.

While "innovation" could be modeled as a top-down process (systemically linking knowledge production to knowledge application), "invention," on the other hand, may be modeled as a bottom-up process. "Creativity" can move top-down as well as a bottom-up (Carayannis & Campbell, 2007, p. 85). The Wikipedia definition of creativity emphasizes the "generation of new ideas or concepts" and "new associations of the creative mind between existing ideas or concepts."[10] *This implies that the creation and production of new knowledge already qualify as examples of and for creativity*. Further propositions are as follows (see again Fig. 2.1): (1) *Creativity in knowledge creation and production is being linked by innovation to knowledge application and use in the wider society.* (2) *Without creativity, the knowledge input for the innovation process might face serious constraints.* (3) *In addition, creativity can also focus on improving processes of innovation on the application and use "side of knowledge."* Creativity management is interested in developing, controlling, regulating, and optimizing creativity for organizations (Dubina, 2005, 2007, 2009). The concept of the "creative knowledge environments" (CKEs) focuses on those environments and contexts that foster creativity in producing new knowledge and new innovations (Hemlin, Allwood, & Martin, 2004).[11] That line of thinking emphasizes to interpret new knowledge as a creative knowledge.[12] Or to rephrase, new knowledge qualifies as a potential candidate for a creative knowledge. This "construction" of creativity as a new knowledge or a new production of knowledge obviously also brings "art" into play. Fiction or science fiction may serve as stimulators for creative ideas, with the potential of being later transformed, at least partially (and of course not always), into new knowledge creation and production. *We can also call this the creativity of knowledge creation.*

The term or concept of innovation can have several meanings. Innovation may mean "change" only or can also refer to an "improvement" or "betterment." In a modern or more recent sense, innovation is being understood mostly as knowledge-based or knowledge-driven. So how can there be a change, improvement, betterment, or reform, which is leveraging, using, and applying knowledge? While knowledge production (or knowledge creation) is often associated closer to research (R&D), innovation expresses a focus of utilizing knowledge for economic (economy), social (society), and political (democracy) purposes. In that sense, mature innovation and innovation systems require a knowledge base or knowledge production.

Bengt-Åke Lundvall paraphrased Boulding (1985) by saying that a system could be seen as the opposite to chaos. In more detail, Lundvall (1992, p. 2) says: "Somewhat more specifically, a system is constituted by a number of elements and by the relationships between these elements." In the words of Kuhlmann (2001, p. 955), a system is: "As a system we understand a conglomeration of actors, institutions and processes all functionally bound together, whereby certain characteristic

[10] See: http://en.wikipedia.org/wiki/Creativity (retrieved: October 29, 2009).

[11] Hemlin et al. provide the following definition for CKEs (quoted from the slide page number 3 of a PowerPoint presentation: http://www.spp.gatech.edu/conference2006/PPTs/Hemlin_7E.pdf, retrieved November 16, 2009), creative knowledge environments = "... those environments, contexts and surroundings, the characteristics of which are such that they exert a positive influence on human beings engaged in creative work aiming to produce new knowledge or innovations, whether they work individually or in teams, within a single organization or in collaboration with others."

[12] Consequently, the influential book *The New Production of Knowledge* (Gibbons et al., 1994) also could have been titled as *The Creative Production of Knowledge*.

core functions of each form the demarcation criteria against other societal (sub) systems." In the process of a definition of a system, often two aspects are coming together: the elements of a system and the self-rational of a system. With the logic of this particular approach, the following definition can be offered for a system: "(1) *Elements*: systems consist of elements (parts); (2) *Self-Rationale*: systems have a mode of operation, a self-rationale (logic, self-logic), which organizes the self-organization and reproduction of a system and the relationship between the elements within a system and, furthermore, the relationship between the system and the other systems" (Campbell, 2001, p. 426; Carayannis, Campbell, & Rehman, 2016, p. 4).

In innovation theory, networks and clusters are important. "Innovation networks" and "knowledge clusters" (Carayannis & Campbell, 2006) introduced here new perspectives. Networks underscore the importance of boundary-transcending interlinking and interlinkages in interdisciplinary, transdisciplinary, and trans-sectoral formats. The concept of sectoral systems of innovation (Malerba, 1999, 2002, 2004) also relates to ideas of clusters and networks. *Smart cities or knowledge cities represent another example for knowledge clusters.* This puts forward the demand and requirement to conceptually bridge (or to "bride" in a metaphorical sense) networks and clusters *with* systems (and systems theory), leading to something like *networks of innovation networks and knowledge clusters:* "One way to look at this, is: clusters could be interpreted as an equivalent for the elements of a system; and networks as a (partial) equivalent for the relationship between the elements of one or of several systems. Networks may represent a specific, but crucial, subset of relations, relationships. *Through networking the clusters/elements of a system (of different systems) relate and interact (and communicate)*" (Carayannis et al., 2016, p. 8). Is a system being embedded by a larger system, then this system qualifies to be interpreted as a cluster or an element of a larger meta-system. Furthermore, elements or clusters within a system could be tested if they also qualify to be considered being a subsystem (or micro-system). This clearly expresses *fractal characteristics* in structure and process.

In a *spatial approach*, the multi-level system approach can address different layers, such as global, supranational, national, regional, and local. The national system of innovation represents here one of the core understandings. Bengt-Åke Lundvall (1992, p. 2) defines the national innovation system in the following way: "It follows that a system of innovation is constituted by elements and relationships which interact in the production, diffusion and use of new, and economically useful, knowledge and that a national system encompasses elements and relationships, either located within or rooted inside the borders of a nation state ... it is obvious that the national system of innovation is a *social* system. A central activity in the system of innovation is learning, and learning is a social activity, which involves interaction between people." In this regard, Lundvall (1992, p. 4) depicted the modern Western nation states as "engines of growth." Also Richard R. Nelson (1990, p. 193) sees capitalism as an "engine of progress" (see also Nelson, 1993). But despite this focus on the national level, Lundvall was from the beginning explicit, by acknowledging also the global but also the regional levels and dimensions of innovation. "Both globalisation and regionalisation might be interpreted as processes which weaken the coherence and importance of national systems" (Lundvall, 1992, p. 3). So, by this, it could be argued that Lundvall had framed his ideas of a national innovation system already from the beginning within the context of multi-level architectures, meaning it does not make sense to talk about a national level without acknowledging global and (subnational) regional and local levels. What Lundvall is saying is that the national level does matter, because a national level is existing. Similarly argues Stefan Kuhlmann. He could be interpreted in a way of suggesting a possible co-evolution between political systems and innovation systems. "Interwoven national and transnational governance mechanisms may feed the development of a transnational political system, including and building upon transformed national systems, fulfilling both 'local' (i.e. regional or national) and 'supra-local' functions at the same time" (Kuhlmann, 2001, p. 956). Also Kuhlmann (2001, p. 954) sees the national level of an innovation system being accompanied in parallel by regional innovation systems: "In the meantime, national and increasingly also regional governments of all these countries pursue, more or less explicitly, 'innovation policies', understood here as the integral of all state initiatives regarding science, education, research, technology policy and industrial modernization, overlapping also with industrial, environmental, labour and social policies."

In that sense, there is also always a momentum of co-evolution between the (multi-level) innovation system and the (multi-level) political system.

Democracy refers to self-government of the people and to basic human rights. A *democracy* can be regarded as a system that is based on four underlying conceptual dimensions: freedom, equality, control, and sustainable development. "In theoretical and conceptual terms, we refer to a Quadruple-Dimensional structure, also a Quadruple Helix structure (a 'Model of Quadruple Helix Structures') of the four basic (conceptual) dimensions of freedom, equality, control, and sustainable development for explaining and comparing democracy and quality of democracy" (Campbell, Carayannis, & Rehman, 2015, p. 467). "There is a potential that democracy discourses and innovation discourses advance in a next-step and two-way mutual cross-reference. The architectures of Quadruple Helix (and Quintuple Helix) innovation systems demand and require the formation of a democracy, implicating that quality of democracy provides for a support and encouragement of innovation and innovation systems, so that quality of democracy and progress of innovation mutually 'Cross-Helix' in a connecting and amplifying mode and manner. This relates research on quality of democracy to research on innovation (innovation systems) and the knowledge economy" (Campbell et al., 2015, p. 468). In a more narrow understanding, a democracy falls together with a "democratic" political system. In a broader understanding, a democracy includes a democratic political system but extends also to the relevant contextualization of the political system. Further attributes of a democracy are pluralism, heterogeneity, and diversity.

The argument here is not that authoritarian (semi-authoritarian) political systems cannot develop a national innovation system. However, the argument is that authoritarian (semi-authoritarian) political systems are not in a position to advance (or to transform) a national innovation system to next higher levels of maturity. Particularly for Russia and China, this is of relevance and will be of further interest in the coming years.

The term and concept of *governance* may be defined as processes of organization or self-organization of different systems, for example, the political, economic, or innovation systems. Governance utilizes strategies and policies in theory and practice.

Analyses of knowledge originally focused more on the knowledge creation and production. Universities and other HEIs were at the core, delivering basic research and educational functions. Innovation as a concept was either not mentioned literally (for example, Bush, 1945) or had a conservative connotation. Joseph Schumpeter, for example, does not make a strong connection to knowledge in his definition of innovation, as is being expressed by the following quote about innovation (taken from Miyata, 2003, p. 715): "(1) an introduction of new products (or products with improved quality); (2) new method of production; (3) new markets and distributing channels; (4) new sources of supply and inputs; (5) new organizations of an industry" (Schumpeter, 1934, p. 66).

Later approaches emphasize the connection of innovation to knowledge by interpreting innovation as a knowledge application, diffusion, and use. Now, innovation is being regarded as essential for the leveraging and "fueling" of knowledge into the society and economy of a knowledge society and a knowledge economy. Innovation carries knowledge far into society and fills all of society with knowledge. Often (not always) this applied knowledge has roots in processes of knowledge creation and production in types and arrangements of basic research. *This inclusion of innovation into the conceptualization of knowledge has the effect that the concept of knowledge is being "broadened" and contextualized by society.* Knowledge is also a social process. Without references to society and social applications and the problem-solving potentials of knowledge, knowledge cannot be understood sufficiently anymore. Knowledge application and use feedback directly into knowledge creation and production (nonlinear innovation models). Concepts such as the national systems of innovation (Lundvall, 1992; Nelson, 1993) or the multi-level innovation systems (Carayannis & Campbell, 2006; Kaiser & Prange, 2004; Kuhlmann, 2001) emphasize these aspects of a society-wide stretch of knowledge. Economic policy is being partially replaced by innovation policy (Carayannis & Campbell, 2006, 2007). *Carrying such ideas consistently further, this also implies that knowledge production and knowledge application, from a systemic perspective, should not only reflect the context of society but, in addition, also the environmental context of society. Knowledge is being contextualized by society but also by the (natural) environments of society.*

Chapter 3
Mode 1, Mode 2, and Mode 3: Triple Helix and Quadruple Helix

The author team of Gibbons, Limoges, Nowotny, Schwartzman, Scott, and Trow (Gibbons et al., 1994)[1] distinguishes between two different modes of knowledge production. "Mode 1" focuses on the traditional role of university research in an elderly "linear model of innovation" understanding. This reflects a basic university research, interested in "first/basic principles" and "discoveries," with a disciplinary research structure, where quality is being controlled primarily by disciplinary peers or a disciplinary peer review process. These disciplinary peers exercise a strong quality gatekeeper function and represent also a university (higher education) system with powerful hierarchies, built into the institutions (Gibbons et al., 1994, p. 1, 3, 24, 33–34, 43–44, 167). Success in Mode 1 (of Mode 1 university research) is defined as a quality or excellence that is approved by hierarchically established peers: "Success in Mode 1 might perhaps be summarily described as excellence by disciplinary peers" (Gibbons et al., 1994). Mode 1 is not concerned with the application, diffusion, and use of knowledge, and Mode 1 does not focus on features in relation to problem-solving for the society or the economy. Nonlinear innovation models are of no major concern for Mode 1.

Mode 2 knowledge production, on the contrary, can be characterized by the following five principles: (1) "knowledge produced in the context of application"; (2) "transdisciplinarity"; (3) "heterogeneity and organizational diversity"; (4) "social accountability and reflexivity"; (5) and "quality control." Mode 2 represents a "problem-solving which is organized around a particular application" and where "Knowledge production becomes diffused throughout society. This is why we also speak of a socially distributed knowledge" (Gibbons et al., 1994 p. 3–4). In Mode 2 the terms "discovery," "application," and "fabrication" (also fabrication of knowledge) overlap. Exploitation of Mode 2 knowledge demands, at least to a certain extent, actual participation in the knowledge production process. Prerequisites of Mode 2 were (are) the massification of tertiary higher education, followed by a

[1] The full names of the whole research team are Michael Gibbons, Camille Limoges, Helga Nowotny, Simon Schwartzman, Peter Scott, and Martin Trow.

E. G. Carayannis, D. F. J. Campbell, *Smart Quintuple Helix Innovation Systems*, SpringerBriefs in Business, https://doi.org/10.1007/978-3-030-01517-6_3

considerable spillover of higher education graduates and higher education knowledge (competencies) into society. Advancing IT technologies allowed an effective communicative linkup of those different knowledge-competent sites outside of the universities and the higher education sector. Continuous communication and negotiations between knowledge producers are crucial. Manifold network arrangements are necessary features for linking together knowledge producing sites "through functioning networks of communication" (Gibbons et al., 1994, p. 6).

The principle of transdisciplinarity underscores the primacy of problem-solving in Mode 2, for which different disciplinary knowledge may be combined or recombined in conventional or unconventional formats. The purity of disciplinary knowledge does not define a criterion of concern. Transdisciplinarity, according to Mode 2, should develop "a distinct but evolving framework to guide problem-solving efforts," is not interested in establishing new academic disciplines, and represents a "problem solving capability on the move" (Gibbons et al., 1994, p. 5). Tacit knowledge (embedded in individual persons or organizations) is as valid or relevant as codified knowledge (written down or stored). In epistemic terms, researchers, in Mode 2, "do not concern themselves with the basic principles of the world but with specific ordered structures within it" (Gibbons et al., 1994, p. 24). Therefore, one may postulate that Mode 2 resembles a transdisciplinary problem-solving knowledge, where "knowledge production in Mode 2 occurs within transient contexts of application" and with "knowledge producers with many different institutional affiliations, either simultaneously or sequentially" (Gibbons et al., 1994, p. 33). Success in Mode 2 means that knowledge was useful or that a knowledge production contributed effectively to a problem-solving in society or the economy: "In Mode 2 success would have to include the additional criteria such as efficiency or usefulness, defined in terms of the contribution the work has made to the overall solution of transdisciplinary problems," and the quality control is being exercised by the "community of practitioners" that do not follow the structure of an institutional logic of academic disciplines (Gibbons et al., 1994, p. 33). Mode 2 demands more social accountability and reflexivity and a greater sensitivity for the impact of knowledge on society and the economy. Values of individuals and of groups must be reflected, to allow social acceptance for a particular problem-solving approach. The authors of Mode 2 (Gibbons et al., 1994) postulate that Mode 2 developed out of Mode 1. Furthermore, there is a parallel existence of Mode 2 and Mode 1 with coevolutionary effects (see generally Gibbons et al., 1994, p. 3, 4, 6–8, 11–14, 19, 29, 33–34, 38, 42, 44–45, 168; see furthermore Campbell, 2006, p. 71–73, 91–92; Campbell & Güttel, 2005, p. 154; Nowotny, Scott, & Gibbons, 2001, 2003).

The "Triple Helix" (three-helix) model focuses on the interaction of the state, academia, and industry. In accordance with the OECD classification of sectors, the state represents the government sector, academia the higher education sector, and industry the business enterprise sector. For Etzkowitz and Leydesdorff (2000, p. 115), the "university-industry-government relations" are of a crucial importance, with universities representing a core institution in the knowledge society: "The university can be expected to remain the core institution of the knowledge sector" (Etzkowitz & Leydesdorff, 2000, p. 117–118). Furthermore: "The Triple Helix thesis states that the university can play an enhanced role in innovation in increasingly knowledge-based societies" (Etzkowitz & Leydesdorff, 2000, p. 109). Research and teaching are central functions of universities. In context of a "second academic revolution," now a "third mission" gains in importance for universities, which assigns to universities also the function of supporting "economic development" (Etzkowitz & Leydesdorff, 2000, p. 110). The US university system after 1945 was guided by the principles of a "peer review" system that allocated funds to a "scientific elite." But the third mission, finally, caused a "breakdown" of this pure peer review system or of the "best science" model, since it linked science to "new sources of legitimating such as regional development," where "science provides much of the basis for future industrial development." The advancing of economic development is being added to the agenda of universities, extending complementarily the original mission of research excellence and teaching. "Less research-intensive regions are by now well aware that science, applied to local resources, is the basis of much of their future potential for economic and social development" (Etzkowitz & Leydesdorff, 2000, p. 116–117). In that context Etzkowitz (2003) also speaks of the "entrepreneurial university." *It appears evident that the so-called third mission displays in substance a series of features similar to the above discussed concept of Mode 2.*

Empirically, different Triple Helix configurations can exist. In the "etatistic model" (a strong state model), the state dominates the other sectors. This may serve as a description for the former communist regimes in the Soviet Union and Eastern Europe. In the "laissez-faire model," the different sectors and institutions are considerably separated. Earlier national systems of innovation in the West, which operated under the premises of linear models of innovation, could represent empirical examples. The "Triple Helix III" model of "trilateral networks and hybrid organizations" of "university-industry-government relations" may be described in the following way: "… is generating a knowledge infrastructure in terms of overlapping institutional spheres, with each taking the role of the other and with hybrid organizations emerging at the interfaces" (Etzkowitz & Leydesdorff, 2000, p. 111–112). According to Etzkowitz and Leydesdorff (2000, p. 112), the Triple Helix III model represents currently for most countries the dominant frame of reference, the crucial benchmark for knowledge and innovation. Key here is the overlap and cross-communication between the different helices or sectors in a knowledge society and economy. In such a context also "nonlinear models of innovation" can be embedded more easily. Some conclusions of Triple Helix are (Etzkowitz & Leydesdorff, 2000, p. 118–119) (1) the nation-state no longer defines the only level for arrangements between government and industrial sectors; (2) profit represents an important driving force; (3) successful innovations change the "landscape," meaning the "opportunity structure" for institutions; (4) the "human capital factor" gains in importance; (5) tensions create a "dynamics for the system," so they do not necessarily have to be resolved; and (6) the communication density within each helix is higher than across the helices; however, in connection to the advancement of systems, the cross-helix communication flow should increase substantially.

Triple Helix, as a model, references explicitly to the models of Mode 1 and Mode 2, by claiming that Mode 2 describes the underlying change in the knowledge production, whereas Triple Helix could be interpreted as an "overlay" at the level of social structures: "The Triple Helix overlay provides a model at the level of social structure for the explanation of Mode 2 as an historically emerging structure for the production of scientific knowledge, and its relation to Mode 1" (Etzkowitz & Leydesdorff, 2000, p. 118; for a summary of Triple Helix see Campbell, 2006, p. 73–74, 92; Campbell & Güttel, 2005, p. 154).

The "Quadruple Helix" (four-helix) model adds to government, universities (higher education), and the economy as further fourth helix the "public," more precisely being defined as the "media-based and culture-based public": "This fourth helix associates with 'media', 'creative industries', 'culture', 'values', 'life styles', and perhaps also the notion of the 'creative class' (a term, coined by Richard Florida, 2004). Plausibility for the explanatory potential of such a fourth helix are that culture and values, on the one hand, and the way how 'public reality' is being constructed and communicated by the media, on the other hand, influence every national innovation system" (Carayannis & Campbell, 2009, p. 206). This fourth helix also could be titled or described as the *media-based*, *culture-based*, and *values-based public*. The Quadruple Helix is analytically broader than the Triple Helix, thus can be used for research questions outside the core focus of Triple Helix. The Quadruple Helix reflects on phenomena such as the "media-based democracy" or a "multimedia information society" (Plasser & Plasser, 2002). Strategies and policies of knowledge and innovation may be supported by communication strategies in or through the media (mass media). Art can be seen as something to foster creativity, implying new forms of knowledge and innovation. Visions in the arts perhaps trigger, in the long run, the development of a new technology or the launch of a next technology cycle. Kuhlmann (2001, p. 954, 958, 967) speaks of "innovation cultures," asserting that a knowledge society and knowledge economy also are being driven by cultures and values. Multicultural settings feed into creativity. The principle of social accountability and reflexivity of Mode 2 has the consequence that the underlying values of individuals, groups, and society as a whole must be recognized and taken into account, so that a knowledge, produced in the context of application and tailored for a problem-solving, is being socially accepted and thus can be successfully applied. Social processes of a knowledge production must be sensitive for culture and the values that influence a society. Here the Mode 2 approach and the Quadruple Helix model interplay.

The concept of "Mode 3" (Carayannis & Campbell, 2006) is being carried by several considerations. For advanced knowledge societies and economies, it is crucial to accept and to foster a pluralism of different knowledge and innovation modes (paradigms). In advanced knowledge societies and economies, this pluralism is being "integrated" on the basis of a coexistence and coevolution of a diversity of knowledge and innovation modes (paradigms), enabling a mutual cross-learning of different *knowledges*. (Over time, some knowledge and innovation modes may become replaced by others.) This makes knowledge more similar to democracy, allowing to speak of a "democracy of knowledge" (Carayannis & Campbell, 2009,

p. 207–208). Key features of Mode 3 are: "Crucial for the suggested 'Mode 3' approach is the idea that an advanced knowledge system may integrate different knowledge modes. Some knowledge (innovation) modes certainly will phase out and stop existing. However, what is important for the broader picture is that in fact a co-evolution, co-development and co-specialization of different knowledge modes emerges. This pluralism of knowledge modes should be regarded as essential for advanced knowledge-based societies and economies. This may point to similar features of advanced knowledge and advanced democracy" (Carayannis & Campbell, 2009, p. 206). "'Mode 3' allows and emphasizes the co-existence and co-evolution of different knowledge and innovation paradigms. In fact, a key hypothesis is: *The competitiveness and superiority of a knowledge system is highly determined by its adaptive capacity to combine and integrate different knowledge and innovation modes* via *co-evolution, co-specialization and co-opetition knowledge stock and flow dynamics* (for example, Mode 1, Mode 2, Triple Helix, linear and non-linear innovation)" (Carayannis & Campbell, 2009, p. 223).[2]

For a multi-level advanced Mode 3 knowledge system, the existence and coevolution of a pluralism and diversity of knowledge and innovation modes are pivotal. This pluralism in fact promises advantages and flexibility and appears necessary for prospects and opportunities in direction of a further development of knowledge societies and knowledge economies. Just as democracy must balance different and opposite viewpoints and is being driven by a pluralistic political spectrum of a variety of political parties, politicians, and voters, also a Mode 3 knowledge society and a Mode 3 knowledge economy require and excel a diversity in knowledge and innovation. This does not rule out that some knowledge or innovation modes can phase out (historically) and are being replaced by other (new) knowledge and innovation modes. The notion of a coevolution of knowledge (and innovation) modes rather emphasizes that despite phenomena of a "paradigm shift,"[3] the general picture of a coexistence of a pluralism of modes is not being questioned (on the "structure of scientific revolutions," see also Kuhn, 1962 and Umpleby, 2005).[4] *Mode 3 encourages interdisciplinary thinking and transdisciplinary application of interdisciplinary knowledge*. Hybrid thinking, parallel and simultaneously in different systems or on the basis of "trans-systemic" conceptual approaches, appears to be key. One could argue that concepts such as "sustainability," "sustainable development," or "social ecology" are already per se interdisciplinary and transdisciplinary, should analysis be followed by application. Research questions and problem-solving in relation to ecology, the environment, environmental changes, and environmental protection increasingly depend in interdisciplinary and transdisciplinary network

[2] On the concept of "co-opetition" (forms or network configurations of a simultaneous cooperation and competition), see Brandenburger and Nalebuff (1997).

[3] Modes of knowledge and innovation may be reinterpreted as "paradigms" or as being paradigm-based.

[4] According to Wikipedia (http://en.wikipedia.org/wiki/Thomas_Kuhn, retrieved: November 12, 2009), the concept of a "paradigm shift" being referred to Kuhn, however, was not literally created by Kuhn.

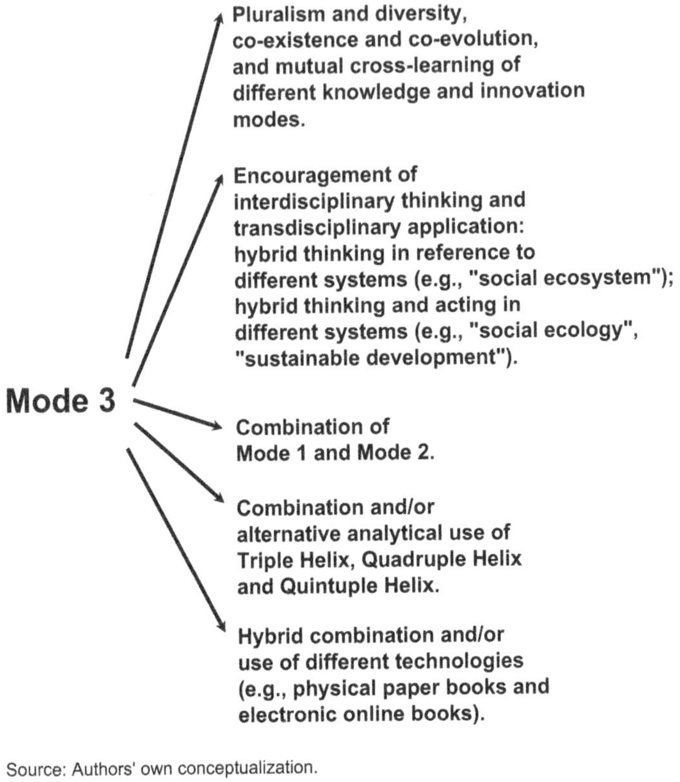

Fig. 3.1 Key features and propositions of Mode 3. Source: Authors' own conceptualization

configurations of different knowledge and innovation modes. Hybridization in Mode 3 also refers to how Mode 1 could be combined with Mode 2 or how Triple Helix may be embedded and contextualized within a wider Quadruple Helix architecture. Hybridization furthermore applies to opportunities of combining different technologies, at least for specific periods: examples for hybrid technologies may be the coexistence of physical paper books in print and electronic (online) books[5] or the coexistence of different drive motors of the coming hybrid and plug-in hybrid cars that most likely will mark a major change for land transportation with hopefully environmentally positive effects such as considerable reductions in CO_2 emissions[6] (see Fig. 3.1).

[5] At least one potential quality of print books will be to serve as a different backup medium (in paper) for the electronic e-books. University libraries again often are challenged of not exactly knowing where to store the masses of print publications in the long run.

[6] Current hybrid cars combine a combustion engine with an electric motor. Next-generation automobiles might be hybrid plug-in hydrogen cars that link an electric motor with a fuel cell. Such cars could either be externally charged directly with electricity or could convert, in the fuel cell, hydrogen and oxygen to electricity (and heat) for the electric motor. Hydrogen cars powered by

Mode 3 claims a certain congruence of structures and processes of advanced knowledge and advanced democracy. In the following two sections ("The Broader Contextualization of Knowledge and the Creation of a Knowledge Democracy" and "The Broadening of the Concept of Democracy and of the Quality of Democracy: Democracy and the Environment"), we want to add some plausibility to these propositions.

The Broader Contextualization of Knowledge and the Creation of a Knowledge Democracy

There are clear indications that the conceptualization and contextualization of knowledge have become increasingly broader. Knowledge creation and production was and still is being extended to knowledge application, diffusion, and use, incorporating ideas of innovation. Knowledge users out in the practical fields are just as important as knowledge producers (knowledge creators), and, depending on the specific constellation or network configuration (e.g., in a nonlinear innovation arrangement), the same person or institution can act as a knowledge producer and/or knowledge user. The combination of Mode 1 and Mode 2 is more extensive than a pure Mode 1 system, and this also holds true for the following combinations: Triple Helix and Quadruple Helix *over* Triple Helix and linear and nonlinear models of innovation *over* one-way linear innovation models. National systems of innovation are being reframed in the context of multi-level systems of innovation. In principle, knowledge for a practical problem-solving of society or the economy has the same relevance as knowledge involved in basic research activities on the fundamental "principles of the world." Transdisciplinarity, here, means the application of interdisciplinary (or also disciplinary) knowledge. This emphasis of the application context of knowledge and the problem-solving interest of innovation imply that "knowledge production becomes diffused throughout society" (Gibbons et al., 1994, p.4). Therefore, in this particular understanding, this form of knowledge represents also a social knowledge.

Empowering citizens as knowledge producers and knowledge users can contribute to a process of "democratizing innovation" (Von Hippel, 2005). Eric von Hippel distinguishes between a "user-centered innovation" and a "manufacturer-centric innovation." The user-centered approach implies that "users of products and services" are "increasingly able to innovate for themselves." "Lead users" are the "innovating users," who can be individuals or firms. *Users innovate so that they*

fuel cells emit only water (water vapor). If the electricity for the plug-in device or the hydrogen for the fuel cell generated in a clean way, this next-generation technology might contribute to a substantial reduction of carbon dioxide emissions of the land-bound traffic and would help balancing the current effects of a global warming of the world climate. Several analysts believe that some of the Japanese and German car companies are (at least for the moment) the global leaders in hydrogen technology.

have what they cannot find on the market. Lead users often "freely reveal their innovations" to others, as being exemplified by the "open source" software movement. "Innovation communities" help to diffuse innovations more quickly. User innovations contribute in general to the social welfare of a society. Manufacturers, in fact, should search for "lead user innovations" and then should consider how these could be re-translated into new products or services, offered by commercial firms. Manufacturers may consider providing "toolkits" with their products or services, so that users can design their own customized solution or application (Von Hippel, 2005, p. 1, 4, 11, 15).[7]

This society-wide stretch of knowledge production and knowledge use implies that knowledge and innovation "flow through" all (at least the major sections) of society: society and the economy are "filled" with knowledge (see Fig. 3.2). When society in general becomes knowledge-based, then this contributes to the establishment of a knowledge-based democracy or even a knowledge democracy. The Mode 3 architecture of knowledge emphasizes that Mode 2-based knowledge for problem-solving often (but of course not always) has hybridized cross-linkages to a Mode 1-based knowledge of basic research in the sciences (in the context of universities), partly in a linear and partly in a nonlinear framework of innovation models. It is evident that widespread knowledge can support democracy and the formation of high-quality democracy. Electoral studies clearly indicate that the higher the level of educational attainment, the more likely a person will vote (for the USA see U.S. Census Bureau, 2008, Table 5). Education thus drives electoral participation rates. Higher education benefits people also in economic and socioeconomic terms (for the USA see Baum & Payea, 2005, and Baum, Payea, & Steele, 2006). Several analyses indicate positive interactions and feedback loops between education, democracy, and the economy (Carmines & Stimson, 1980; Saint-Paul & Verdier, 1993). Values are sometimes being typologized and contrasted in the two groups of "materialist" and "postmaterialist" values. Postmaterialism is more sensitive for environmental issues. There is a hope that economic progress finally gives rise to postmaterialist values in the long run: "The scarcity hypothesis implies short-term changes, or period effects: Periods of prosperity lead to increased Postmaterialism, and periods of scarcity lead to Materialism. The socialization hypothesis implies that long-term cohort effects also exist: the values of a given generation tend to reflect the conditions prevailing during its preadult years" (Inglehart, 1990, p. 75, 79). Should values diffuse and become more dominant in favor of a greater protection of the environment, then a problem-solving in Mode 2, which demands more social accountability and reflexivity, would have to recognize such a value shift. This would increase opportunities for "eco-innovation" and "eco-entrepreneurship."

[7]Two key books of Eric von Hippel, *The Sources of Innovation* (1988) and *Democratizing Innovation* (2005), are electronically available as a free download (http://web.mit.edu/evhippel/www/books.htm). Print versions must be purchased. This illustrates how a print medium and an electronic medium of the same publication can be combined in an innovative, creative, and effective way and furthermore might indicate a promising hybrid strategy for publishers in the future.

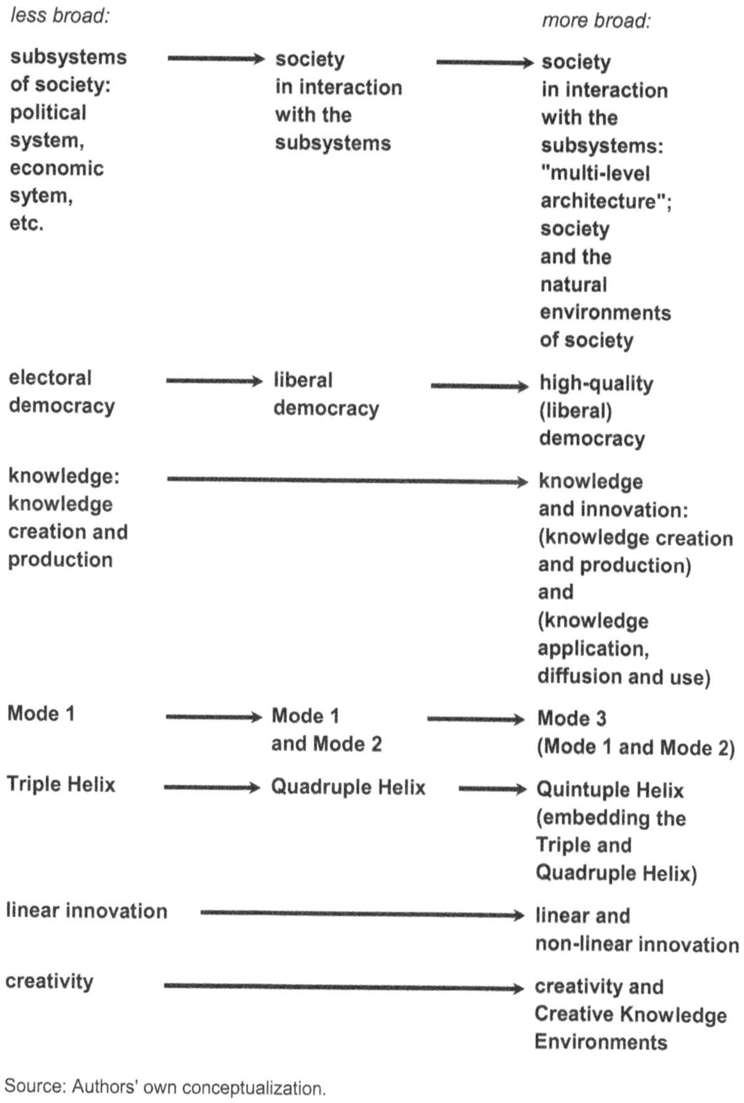

Fig. 3.2 Trends in the broadening of concepts. Source: Authors' own conceptualization

Roeland J. in't Veld (2010a, 2010b) developed and proposed an advanced and mature concept, how the structures and dynamics of and within a knowledge democracy can be framed further. He has put a particular emphasis on roles and responsibilities of the media. For him, there are three operating crucial forces as crucial key references: "emerging participatory democracy," "emerging transdisciplinary design/science," and "emerging bottom-up media" (in't Veld, 2010b, p. 11). The bottom-up media are being complemented by the more "top-down media." On the relationship of media an politics, in't Veld (2010b, p. 4) is providing the following assessment: in his opinion he says that the "Media and politics [are], a relationship based on mutual interest as on the other hand the media equally need politicians in order to produce news, one of their main products. So this dependence is reciprocal."

The Broadening of the Concept of Democracy and of the Quality of Democracy: Democracy and the Environment

In congruence to a tendency that knowledge has been conceptualized more broadly over time, by extending knowledge from knowledge creation and production to knowledge application, diffusion, and use (furthermore emphasizing a pluralism of knowledge modes, thus the metaphor of a "democracy of knowledge"), one can formulate the proposition that there is also a tendency that the concepts of democracy increased their complexity. Minimum definitions of democracy are being challenged by maximum definitions. Originally, democracies were described in terms of an "electoral democracy," focusing on political rights and on issues of elections. Robert A. Dahl (1971, p. 2–9) explains democracy as the interplay of the two dimensions of "public contestation" ("political competition") and "participation." The "liberal democracy" already is more demanding than a pure electoral democracy, adding to the political rights the civil liberties. The country-based freedom measures, in a global comparative format, produced by Freedom House (2008, 2009a, 2009b), refer to such a liberal democracy understanding, since their measures focus on and combine political rights and civil liberties. In modern democratic theory, the originally two dimensions of democracy of Dahl have been substantially complemented. In a review about the quality of democracy, Larry Diamond and Leonardo Morlino (2004, p. 22–23) identify the following eight dimensions that appear crucial for a democracy and the quality of democracy: rule of law, participation, competition, vertical accountability, horizontal accountability, freedom, equality, and responsiveness.

Another question is whether democracy represents a concept only of the political system or, alternatively, a concept that extends to society and thus also focuses on the interfaces of the political system with society and the economy. For Guillermo O'Donnell (2004, p. 13), the human beings (as "agents") are endowed with the following characteristics: they have (in principle) the autonomy to make decisions; they have the cognitive ability to reason; and they have a responsibility for their own actions. Already at this point, it appears to be evident why people (human beings), in a society enriched with knowledge and never-ending knowledge flows, are better prepared to act as conscious "agents" who reflect their democracy politically and who are engaged in a political decision-making. O'Donnell (2004, p. 12–13, 42, 47) defines the following two key dimensions for democracy and the quality of democracy: "human rights" (e.g., political rights, civil rights, and social rights) and "human development." O'Donnell (2004, p. 55) uses the metaphor of a "nexus of these three currents," where *democracy*, *human rights*, and *human development* are intertwined. The conceptual formula of O'Donnell for the quality of democracy thus may be paraphrased as (Campbell, 2008, p. 41): "quality of democracy = (human rights) + (human development)." By incorporating human development, O'Donnell (2004, p. 11–12) carries his understanding of democracy and the quality of democracy already far out into society, because he draws a direct intellectual line to the Human Development Reports and the Human Development Index (HDI), which is being regularly and annually released by the United Nations Development Program (UNDP). Interpreting O'Donnell freely and referring to his approach as a theoretical point of departure, one could set up the hypothesis that, at least in principle, the HDI qualifies as a measure for human development in a comparative global format (see, e.g., UNDP, 2009, p. 171–175). O'Donnell emphasizes that human development actually transforms the human rights from rights into real freedoms.[8]

The "Democracy Ranking of the Quality Ranking" applies, as underlying model of democracy, a broad conceptualization of democracy and the quality of democracy, which is even more encompassing than the approach of O'Donnell.[9] The conceptual formula of the "Democracy Ranking" is "quality of democracy = (freedom and other characteristics of the political system) + (performance of the non-political dimensions)." In addition to the political system, the performance of the non-political dimensions also is being factored in. With this focus on performance, the "Democracy Ranking" attempts to be "neutral" with regard to a left/right or liberal/conservative axis, not favoring either left or right values, ideologies, or policies, but looks more closely on the output of performance that should be empirically accessible and indicator-based for reasons of measurement. In Western political thought, traditionally, freedom often is more closely associated with the right or conservatism and equality with the left (Harding, Phillips, & Fogarty, 1986, p. 87). The non-political dimensions, in the context of the "Democracy Ranking," are gender, economy, knowledge, health, and the environment (Campbell, 2008, p. 30–41; Campbell & Barth, 2009, p. 216–218; Campbell & Pölzlbauer, 2009, p. 3–8; Campbell & Sükösd, 2002).

For the "Democracy Ranking," the concept of democracy goes beyond the boundaries of the political system and includes the intersections between politics and society, but also the performance of society, which is being interpreted as a responsibility of politics. The "Democracy Ranking" reflects also explicitly on the embeddedness of society in the context of the natural environment (environments), more directly of course the impact of society on nature. Environmentally sensitive behavior of people and society would factor into the "Democracy Ranking" as a good-quality environment. O'Donnell (see again 2004, p. 55) refers to the three-current understanding of democracy, human rights, and human development.

---

[8] As an interesting example for a citizen audit on the quality of democracy, which was carried out in recent years, see Cullell (2004) on Costa Rica.

[9] The general website address of the "Democracy Ranking" is http://www.democracyranking.org/en

The "Democracy Ranking" applies a four-current understanding that links together democracy, human rights, human development, and the (natural) environment of society. Here an interplay is being constructed between the quality of democracy and the quality of the environment. Those cross-references between the political system, the society, the economic system, and the environment indicate that the "Democracy Ranking" model reveals socio-ecological features of sustainable development. While the industrialized nations or the advanced OECD countries often rank high with regard to the quality of their human rights or their economic and socioeconomic performance, they often also pollute the environment considerably more than many of the so-called less or least developed countries (LDCs, LLDCs). In an age of a growing importance of global interwovenness and global responsibility, this for a large part negative impact of the industrialized countries on the environment should be taken more strictly into account. In the "Human Development Report 2007/2008" (UNDP, 2007, p. 21–47), devoted most importantly to the issue of fighting climate change, clearly a link is being drawn between increased CO_2 emissions (and other greenhouse gases) and rising temperatures. The "world is warming" because of "human-induced climate change." Above all the industrialized countries and regions cause most of the global CO_2 emissions; thus they express a negative balance of "deep carbon footprints" (on the concept of the "ecological footprint," see also Monfreda, Wackernagel, & Deumling, 2004).

There are different initiatives, interested in measuring the quality of the environment. For example, the "Environmental Sustainability Index" (ESI) focuses on the "ability of nations to protect the environment over the next several decades." For that purpose 76 different data sets were aggregated into 21 "indicators of environmental sustainability," referring to the following features: "natural resource endowments, past and present pollution levels, environmental management efforts, and the capacity of a society to improve its environmental performance" (Esty, Levy, Srebotnjak, & de Sherbinin, 2005, p. 1). The Environmental Sustainability Index was published for the last time for 2005. The follow-up product is the Environmental Performance Index (EPI), which, so far, was released for 2006 and 2008. The EPI framework focuses on offering a "composite index of current national environmental protection efforts." There are two key core objectives: "reducing environmental stresses to human health (the Environmental Health objective)" and "protecting ecosystems and natural resources (the Ecosystem Vitality objective)." For that purpose the EPI applies 25 indicators that are being aggregated to 3 distinct levels (policy categories, objectives, and the final index) (Esty et al., 2008, p. 10, 15–34).[10]

Schumpeter's concept of the built-in "creative destruction" mechanism of a capitalist economy can be explained, in a modern knowledge-based language, with the need of managing simultaneously different technology life cycles and the conversion from "old" to "new" technology life cycles (on technology life cycles, see Campbell, 2006, p. 74–75, 92–93; Tassey, 2001). Technology life cycles link "knowledge waves" to the growth (growth and decline) cycles and long-term performance and competitiveness of an economy. Technology life cycles drive an

[10] See also: http://epi.yale.edu/Home

economy and demand permanent change. Every technology life cycle has an "expiration date," but always new technology cycles are being created. Several technology life cycles, at different stages of market maturity, operate in parallel. Therefore, innovation and innovativeness represent crucial characteristics of firms in a market economy. Economic performance depends on entrepreneurs, who leverage the momentum and dynamics of technology life cycles.[11]

Schumpeter (1976/1942, p. 82–83) provides the following famous quote on the *creative destruction:* "Capitalism, then, is by nature a form or method of economic change and not only never is but never can be stationary. And this evolutionary character of the capitalist process is not merely due to the fact that economic life goes on in a social and natural environment which changes and by its change alters the data of economic action; this fact is important and these changes (wars, revolutions and so on) often condition industrial change, but they are not its prime movers. Nor is the evolutionary character due to a quasi-automatic increase in population and capital or to the vagaries of monetary systems of which exactly the same thing holds true. The fundamental impulse that sets and keeps the capitalist engine in motion comes from the new consumers' goods, the new methods of production or transportation, the new markets, the new forms of industrial organization that capitalist enterprise creates. ... This process of Creative Destruction is the essential fact about capitalism."

The concept of Mode 3 is more inclined to emphasize the coexistence and coevolution of different knowledge and innovation modes. Mode 3 even accentuates such a pluralism and diversity of knowledge and innovation modes as being necessary for advancing societies and economies. This pluralism supports processes of a mutual cross-learning from the different knowledge modes. Between Mode 1 and Mode 2, manifold creative arrangements and configurations are possible, linking together basic research and problem-solving. Individual knowledge and innovation modes may phase out and become replaced in the context of a "paradigm shift" (see again Kuhn, 1962). There also may be some cyclical patterns, indicating how dominant or non-dominant certain modes are during certain periods, captured by the phrase of "knowledge swings" (Carayannis & Campbell, 2009, p. 225). This, however, does not alter the general pattern of a coexistence and coevolution of a continuous and continuing diversity of knowledge and innovation modes. The Quadruple Helix model adds to the "university-industry-government relations" the fourth helix of a "media-based and culture-based public" that also includes values and different value systems.

For the advanced knowledge societies and knowledge economies, we can set up for discussion the following propositions about possible "evolutionary" effects

[11] For an analysis of the different dynamics in the biotechnology and ICT sectors in Finland, Christopher Palmberg and Terttu Luukkonen (2006, pp. 160–161, 167–169) apply the concept of the "competence block." Here the "entrepreneur" is crucial. Palmberg and Luukkonen define the entrepreneur as: "*Entrepreneurs*, or innovators, who *identify* profitable inventions and introduce them in the market. The task of the *entrepreneur* is to identify those ideas that have the greatest potential commercial value and therefore to contribute to turning inventions into innovations in the market."

(described by the concepts of Mode 3 and Quadruple Helix): (1) the pluralism of the knowledge and innovation modes suggests features similar to and in congruence with the political pluralism and diversity of democracy. The notion of a "democracy of knowledge" (Carayannis & Campbell, 2009, p. 207–208) describes these phenomena. (2) The hybrid coupling of Mode 1 basic research and Mode 2 problem-solving leads to a society-wide diffusion of good-quality knowledge. Knowledge is being broadly contextualized by society. Innovation carries knowledge application, diffusion, and use far out into society and the economy. Knowledge producers and knowledge users are cross-linked in heterogeneous networks, with shifting functions and continuous reconfigurations. The same persons and institutions can act simultaneously as knowledge producers and knowledge users. This society-wide flow of knowledge (claimed by Mode 3 and Quadruple Helix) also supports citizenry and political citizenship for a high-quality democracy. Here knowledge society, knowledge economy, and knowledge democracy meet and overlap. (3) Over time, concepts of democracy have become more complex and demanding. Broader conceptualizations of democracy transcend the boundaries of the political system and integrate the interplay of politics, society, and economy. In such a wider understanding, the coevolution of human rights and human development is crucial (e.g., see O'Donnell, 2004). A next step in broadening the concept of democracy would be to blend together the coevolution of human rights, of human development, and of the environment. Cross-linking human rights, human development, and the environment already bridges analytically into sustainable development, clearly including features of social ecology.

It is not easy to balance Schumpeter's concept of "creative destruction," contextualized in a modern interpretation in the framework of the technology life cycles, with the pluralism and coevolution approach of Mode 3 and the Quadruple Helix. Of course one could attempt to juxtapose the two spheres of (1) pluralistic knowledge and innovation modes *and* (2) the dynamics of the technology life cycles, where technology life cycles depart from specific knowledge and innovation modes. But some conceptual tensions between these two different understandings still remain. Schumpeter's model emphasizes more the aspects of competition or of a radically competitive capitalist economic system. Mode 3 and Quadruple Helix are more in favor of stressing the opportunities of coevolutionary learning. In that sense Mode 3 and Quadruple Helix indicate a path of sustainable development for an economic system, interested in advancing a market economy that is socially and environmentally sensitive, thus recognizing and implementing criteria of "social ecology." Here is sufficient space and are sufficient opportunities for "eco-innovation" and "eco-entrepreneurship." Mode 3 and Quadruple Helix may help in converting the "creative destruction" (at least partially) into a "creative learning" and a "creative coevolution."

Chapter 4
Sustainable Development, Social Ecology, and the Quintuple Helix

Society could be designed or understood to consist of different subsystems (or systems).[1] The political system and the economic system are such examples. Politics and the economy are being embedded by society; thus society, in this understanding, is more comprehensive than politics and the economy. For every societal subsystem, the other subsystems of society or society as a whole represents "social environments" (societal environments). In a spatial (spatial-political) multi-level architecture, societies could be located at different levels of aggregation, ranging from sub-national (local, regional) to national and transnational (supranational, global). Society again is being contextualized by the "natural environment" (the natural environments).

In everyday language, when not further specified, the term environment normally is being associated with the natural environment. The planet Earth has a natural environment. The concept of a natural environment may also be applied to other planets (or moons). "Ecology" refers to the interdisciplinary analysis of either interactions between living organisms or interactions between living organisms and their environments. Based on those interaction patterns, the sum of living organisms and of the nonliving environment defines an "ecosystem."[2] "Sustainability" can focus either on the relationship of society to the economy (e.g., socioeconomic regimes or configurations) or the relationship of society with the natural environments (Adams, 2006, pp. 1–3; Winiwarter & Knoll, 2007, pp. 306–307). Concerning biological systems, "biodiversity" represents an indicator for sustainability (see Vadrot, 2008, pp. 62–79). The Human Development Index of the Human Development Reports (UNDP, 2007, 2009) can be interpreted as a measure of "sustainable development" of societies or of countries in global comparison. A key quote on sustainable development pinpoints on a definition of the so-called Brundtland Commission that states

[1] A system could be defined as consisting of "elements/parts" and the "rationale/self-rationale" of these system elements (Carayannis and Campbell, 2009, p. 204). Of course, alternative definitions for a system also are possible.

[2] See also: http://en.wikipedia.org/wiki/Ecology (retrieved, November 06, 2009).

E. G. Carayannis, D. F. J. Campbell, *Smart Quintuple Helix Innovation Systems*, SpringerBriefs in Business, https://doi.org/10.1007/978-3-030-01517-6_4

that sustainable development "meets the needs of the present without compromising the ability of future generations to meet their own needs" (United Nations, 1987a, 1987b; see also Winiwarter & Knoll, 2007, p. 305).

Referring to our argumentation in the previous sections of the analysis, we could also define sustainable development as a coevolution of the different systems of society, based on knowledge and a mutual cross-learning that is socially and environmentally sensitive and that is receptive for concepts of a quality of democracy. "Social ecology" looks at the "society-nature interactions" between "human society" ("culture," the "cultural (symbolic) sphere of causation") and the "material world" ("nature," the "natural (biophysical) sphere of causation"). The "biophysical structures" or "biophysical structures of society" mark an area of overlap between culture (the cultural) and nature (the natural), and between these "biophysical structures" and nature, a metabolism (or a "social metabolism," with potential of a "socio-metabolic transition"), in context of specific "metabolic profiles," occurs (see Fischer-Kowalski, 1998; Fischer-Kowalski & Haberl, 2007; Fischer-Kowalski & Hüttler, 1999; Haberl, Fischer-Kowalski, Krausmann, Weisz, & Winiwarter, 2004, pp. 201–202, 204, 2009; see also Hopwood, Mellor, & O'Brien, 2005; Kates et al., 2001).[3] "Sociometabolic regimes represent dynamic equilibria of society-nature interactions and are characterized by typical patterns of material and energy flows (metabolic profiles)" (Krausmann, Fischer-Kowalski, Schandl, & Eisenmenger, 2008, p. 1). Sustainable development and social ecology represent areas and fields for interdisciplinary analysis and transdisciplinary problem-solving. "Sustainability science is emerging as a transdisciplinary effort to come to grips with the much-needed symbiosis between human activity and the environment" (Rapport, 2007, p. 77).

The originally natural science-based biological concept of the "ecosystem" may also be reinterpreted by the social sciences and redesigned to fit the purpose of a "social or societal ecosystem." A societal ecosystem would embed the crucial "elements" (e.g., actors, institutions, structures, and processes) as well as their complex interaction patterns that characterize an ecosystem but would also stretch into the contextualization by the social (societal) environments of the other systems (subsystems) of society and is finally contextualized by the natural environment of the whole society. A societal ecosystem also (at least potentially) interacts with its social and natural environments. An example for a societal ecosystem would be the "innovation ecosystem" that focuses on the complexity of innovation and innovation systems, framed by societal and natural environments (Carayannis & Campbell, 2009, pp. 201–203, 206, 208; see, furthermore, Milbergs, 2004, pp. 5, 8, 13, 2005a, 2005b, p. 8). For the "innovation ecosystem," the nonlinear models of innovation are of a key importance. The concepts of "biological ecosystems" and of "social ecosystems" (societal ecosystems) demonstrate the whole interdisciplinary stretch of "ecology," underpinning the intellectual and academic challenge of cross-referring and cross-relating ideas between the social sciences and natural sciences, but also highlight the benefit of interdisciplinary inquiry for transdisciplinary

[3] See also: http://www.uni-klu.ac.at/socec/inhalt/1860.htm.

application. Social ecosystems and biological ecosystems could be covered and integrated by a transdisciplinary framework based on "social ecology."

When the relationship and interplay of society and the economy are being regarded as a (possible) criterion for sustainable development, then it appears plausible that Guillermo O'Donnell's (e.g., 2004) conceptualization of the quality of democracy, tying together and integrating human rights and human development, also qualifies as a sustainable development approach. We could claim here an overlap (at least partial overlap) between the concepts of the quality of democracy and of sustainable development. *Is sustainable development a route to high-quality democracy? Or does the quality of a democracy manifest itself in patterns of sustainable development?* Broader conceptualizations or definitions of democracy that do not limit democracy to the political system but are interested in integrating the political system, the society, and the economy in the one or other configuration and under the "umbrella" of democracy potentially reflect aspects of sustainable development. Between the so-called maximum definitions of democracy and sustainable development manifold, theoretical windows of congruence open up. Should a conceptualization of a democracy or the quality of a democracy be designed so broadly as to reflect also the (natural) environmental context of society, then such a framing would not only be compatible with a sustainable development framework in general but would also incorporate features of "social ecology." *Therefore, a concept of knowledge democracy (quality of democracy) that links together the political system, the society, the economy, and the environment allows the application of concepts of social ecology in a framework of sustainable development.*

In this analysis, so far, we arrived at the following conclusions or suggested the following propositions for discussion:

1. *The broadening and "societal contextualization" of the concept of knowledge and of knowledge by incorporating innovation:* Traditional concepts of knowledge focused more on knowledge creation and production, for example, basic university research in the context of higher education systems. Later concepts also included knowledge application, diffusion, and use, emphasizing that innovation could be regarded as using knowledge for application and problem-solving. Innovation-oriented knowledge diffused and still diffuses far out into society and is being characterized as a "social" ("societal") knowledge, contextualized by society. Key in that context is also the concept of the "national system of innovation" (Lundvall, 1992; Nelson, 1993). The whole spectrum of knowledge stretches from the creation and production of new knowledge to innovation and the application and use of knowledge, frequently in nonlinear models of innovation. Creativity refers either to new knowledge or to new innovation. Interestingly, for the global level of innovation systems, Lundvall (1992, p. 7) claims that noneconomic aspects, such as "ecological sustainability" and a reduction of "extreme social inequality," gain importance. In context of this broadening and society-wide stretch of knowledge, two theories (models) on knowledge and innovation are pivotal: Triple Helix (Etzkowitz & Leydesdorff, 2000) looks at the dynamic interaction of the "helices" of "university-industry-government

relations." In the Mode 1 and Mode 2 approach (Gibbons et al., 1994), the basic university research (Mode 1) is being supplemented by a knowledge (Mode 2) that focuses on a problem-solving for the society and the economy.

2. *A possible (partial) congruence and coevolution of knowledge and democracy:* Here, two developments run in parallel that have features of a congruence and coevolution.

   - *The pluralization of knowledge:* Advanced and further advancing (multi-level) knowledge and innovation systems can be characterized by a pluralism and diversity of knowledge and innovation modes. This pluralism is in fact necessary for promoting the continued development of knowledge societies and knowledge economies. Based on such a dynamics, a "democracy of knowledge" emerges, with pluralistic knowledge and innovation modes, with possible coevolutionary effects of a cross-learning. Advanced knowledge takes over structural elements of a democracy, i.e., behaves like a pluralistic democracy. "Mode 3" (Carayannis & Campbell, 2006, 2009) emphasizes this pluralism and coevolution of different and diverse knowledge and innovation modes.[4] Cross-learning between knowledge modes in Mode 3 potentially softens the sharp edges of the "creative destruction" in the economic-technological vision of Schumpeter (1976/1942) and moves the systems in favor of a "creative learning" and a "creative coevolution." Mode 3 stresses hybrid combinations and possibilities of combination between Mode 1 and Mode 2 or between basic research, on the one hand, and applied research and experimental, on the other.[5] For Mode 3 it is crucial that Mode 2 problem-solving in the twentieth and twenty-first centuries (and most likely also beyond) is cross-connected with types of a Mode 1 basic research.[6] "Quadruple Helix" (Carayannis & Campbell, 2009) adds to the "university-industry-government relations" of the Triple Helix model the fourth helix of a "media-based and culture-based public" that also includes culture and values. This spreading of knowledge also helps building a knowledge democracy with political citizens that have the knowledge of making informed decisions.
   - *The broadening of democracy:* Theories of democracy have become increasingly complex over time. Concepts on liberal democracy are more demanding than the simpler versions of an electoral democracy. Some approaches emphasize that democracy is not just a description of the political system but

[4] Government/opposition cycles of the political system find a partial equivalent in the so-called concept of "knowledge swings," referring to the possibility of a sequential patterning of which modes of knowledge or innovation are dominant in which periods of time (Carayannis and Campbell, 2009, p. 225).

[5] This also leads to the question whether Mode 3 encourages that professionals carry hybrid competences and thus qualify as "polyvalence professionals" (see, e.g., Meglic et al., 2009).

[6] In earlier historical periods, also variations of a Mode 2 problem-solving existed, but with less or no cross-connections to a science-based Mode 1 knowledge. In that line of thinking, Mode 2 might be "older" than Mode 1 (see also Etzkowitz and Leydesdorff, 2000, p. 116).

also cross-refers to society, the economy, and other subsystems of society. O'Donnell (e.g., 2004) defines the quality of democracy out of an interplay of "human rights" and "human development." The "Democracy Ranking" model of quality of democracy (e.g., Campbell, 2008) goes even further, adding also the (natural) environment or the support of the natural environment to its conceptualization. Where a model of democracy crosscuts human rights, human development, and environmental development, there are clearly references to "social ecology." A high-quality democracy is more complex than a medium-quality liberal democracy. High-quality democracies depend on a pluralized and advanced knowledge and innovation to perform. *The diversity of a democracy obviously supports the diversity of knowledge. Here the new complexity of knowledge and of democracy meets and comes together.* High-quality democracy is a knowledge-based democracy, a knowledge democracy.

How do knowledge, innovation, and the environment (natural environment) relate to each other? Societies or democracies (high-quality democracies), based on a coevolution of the subsystems of society or of the subsystems in interaction with the whole of society, where mutual learning and a "positive" learning interaction take place, follow the rationale of sustainable development. Advanced and pluralized knowledge, with a coevolution and mutual learning processes between different knowledge and innovation modes, also adopts the rationale of sustainable development. For the purpose of further discussion and analysis, we lastly want to propose and introduce the five-helix model of the "Quintuple Helix," where the environment or the natural environments represent the fifth helix (see Fig. 4.1). The Triple Helix focuses on "university-industry-government relations." The Quadruple Helix frames the Triple Helix in context of a "media-based and culture-based public." The Quintuple Helix finally embeds the Quadruple Helix (and the Triple Helix) in context of the environment or the natural environments.[7] Depending on the analytical point of departure or on the practical interest of application and decision-making, either a Triple Helix, a Quadruple Helix, or a Quintuple Helix model could be more appropriate.

The Quintuple Helix model is interdisciplinary and transdisciplinary at the same time: the complexity of the five-helix structure implies that a full analytical understanding of all helices requires the continuous involvement of the whole disciplinary spectrum, ranging from the natural sciences (because of the natural environment) to the social sciences and humanities (because of society, democracy, and the economy). The Quintuple Helix also is transdisciplinary, since it can be used as a frame of reference for decision-making in connection to knowledge, innovation, and the (natural) environment. The Quintuple Helix can be proposed as a framework for transdisciplinary (and interdisciplinary) analysis of sustainable development and

[7] At this point we leave it open, what in the logical continuation of such a conceptual sequence a Sextuple Helix (six-helix model) or a Septuple Helix (seven-helix model) possibly may or could be.

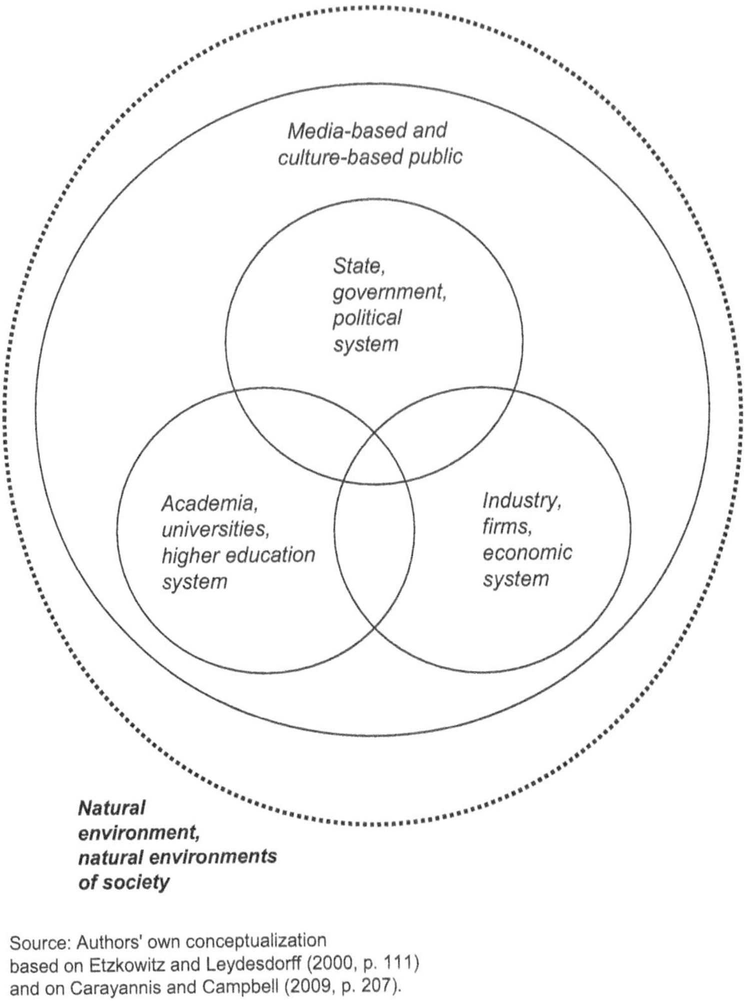

Fig. 4.1 The five-helix model of the Quintuple Helix. Source: Authors' own conceptualization based on Etzkowitz and Leydesdorff (2000, p. 111) and on Carayannis and Campbell (2009, p. 207)

social ecology. With the adding of the "fifth helix of the (natural) environment/environments" to knowledge creation, production, application, diffusion, and use, knowledge and innovation (advanced and pluralized Mode 3 knowledge and innovation systems) are transformed to a knowledge and innovation that is sensitive or at least potentially sensitive for "social ecology": knowledge and innovation, contextualized by society, meets the context of society, the environment. *Therefore, the Quintuple Helix has the potential to serve as an analytical framework for sustainable development and social ecology, by conceptually relating knowledge and innovation to the environment.* Sustainable knowledge is a knowledge that reflects on the performance and quality of the environment, the natural environment. The Quintuple Helix furthermore outlines what sustainable development might mean and imply for "eco-innovation" and "eco-entrepreneurship" in the current situation and for our future.

Chapter 5
Innovation Systems in Conceptual Evolution: Mode 3 Knowledge Production in Quadruple and Quintuple Helix Innovation Systems

Triple Helix Innovation Systems and Mode 1 and Mode 2 of Knowledge Production

Universities, or higher education institutions (HEIs) in more general, have three main functions: teaching and education, research (research and experimental development, R&D), and the so-called "third mission" activities, for example, innovation (Campbell & Carayannis, 2013b, p. 5). In reference to "arts universities" now, the question and challenge arise, whether to which extent and in which way the arts universities differ from the (more traditional) universities in the sciences. Arts universities obviously place an emphasis on the arts, and the arts are not identical with the sciences. However, also arts universities frequently make references to the sciences; thus also arts universities can express competences in teaching and in carrying out research in the sciences. *The other major challenge of arts universities is to engage in "artistic research" and "arts-based innovation." By this, arts universities (and other higher education institutions in the arts) are also being linked to and are being interlinked with national innovation systems and multilevel innovation systems.* This widens the whole interdisciplinary and transdisciplinary spectrum of higher education systems. *"Artistic research" furthermore complements the "teaching of arts" at arts universities* (see also the propositions formulated by Bast, 2013). Hybrid and innovative combinations of universities of arts and universities of the sciences are possible and indicate organizational opportunities for promoting creativity (Campbell, 2013b).

University research, in a traditional understanding and in reference to universities in the sciences, focuses on basic research, often framed within a matrix of academic disciplines, and without a particular interest in the practical use of knowledge and innovation. This model of university-based knowledge production also is being called "Mode 1" of *knowledge production* (Gibbons et al., 1994). Mode 1 is also compatible with the linear model of innovation, which is often being referred to Vannevar Bush (1945). The linear model of innovation asserts that first there is basic

E. G. Carayannis, D. F. J. Campbell, *Smart Quintuple Helix Innovation Systems*, SpringerBriefs in Business, https://doi.org/10.1007/978-3-030-01517-6_5

research in university context: gradually, this university research will diffuse out into society and the economy. It is then the economy and the firms that pick up the lines of university research and develop these further into knowledge application and innovation, for the purpose of creating economic and commercial success in the markets outside of the higher education system. Within the frame of linear innovation, there is a sequential "first-then" relationship between basic research (knowledge production) and innovation (knowledge application).

The Mode 1-based understanding of knowledge production has been challenged by the new concept of "Mode 2" of knowledge production, which was developed and proposed by Michael Gibbons et al. (1994, p. 3–8, 167). Mode 2 emphasizes a knowledge application and a knowledge-based problem-solving that involves and encourages the following principles: "knowledge produced in the context of application," "transdisciplinarity," "heterogeneity and organizational diversity," "social accountability and reflexivity," and "quality control" (see furthermore Nowotny et al., 2001, 2003, 2006). Key in this setting is the focus on a knowledge production in contexts of application. Mode 2 expresses and encourages clear references to innovation and innovation models. The linear model of innovation also has become challenged by nonlinear models of innovation, which are interested in drawing more direct connections between knowledge production and knowledge application, where basic research and innovation are being coupled together not in a first-then but in an "as well as" and "parallel" (parallelized) relationship (Campbell & Carayannis, 2012). Mode 2 appears also to be compatible with nonlinear innovation and its ramifications.

The Triple Helix model of knowledge, innovation, and university-industry-government relations, which was introduced and developed by Henry Etzkowitz and Loet Leydesdorff (2000, p. 111–112), asserts a basic core model for knowledge production and innovation, where three "helices" intertwine, by this creating a national innovation system. The three helices are identified by the following systems or sectors: academia (universities), industry (business), and state (government). In the current innovation discourses, the "Triple Helix" model represents something like a "standard model" of (and for) innovation (by this being something like a "null hypothesis"). Etzkowitz and Leydesdorff refer to "university-industry-government relations" and networks, putting, a particular, emphasis on "trilateral networks and hybrid organizations," where those helices overlap in a hybrid fashion. Etzkowitz and Leydesdorff (2000, p. 118) also explain, how, in their view, the Triple Helix model relates to Mode 2: the "Triple Helix overlay provides a model at the level of social structure for the explanation of Mode 2 as a historically emerging structure for the production of scientific knowledge and its relation to Mode 1." More recently, Leydesdorff (2012) also introduced the notion of "N-Tuple of helices" (Park, 2014).

Quadruple and Quintuple Helix Innovation Systems and Mode 3 of Knowledge Production

Mode 1 and Mode 2 may be characterized as "knowledge paradigms" that underlie the knowledge production (to a certain extent also the knowledge application) of higher education institutions and university systems. Success or quality, in accordance with Mode 1, may be defined as "academic excellence, which is a comprehensive explanation of the world (and of society) on the basis of 'basic principles' or 'first principles', as is being judged by knowledge producer communities (academic communities structured according to a disciplinary framed peer review system)." Consequently, success and quality, in accordance with Mode 2, can be defined as "problem-solving, which is a useful (efficient, effective) problem-solving for the world (and for society), as is being judged by knowledge producer and knowledge user communities" (Campbell & Carayannis, 2013b, p. 32; see furthermore Campbell & Carayannis, 2013c, 2016a). A "Mode 3" university, higher education institution, or higher education system would represent a type of organization or system that seeks creative ways of combining and integrating different principles of knowledge production and knowledge application (e.g., Mode 1 and Mode 2), by this encouraging diversity and heterogeneity and by this also creating creative and innovative organizational contexts for research and innovation (Carayannis & Campbell, 2006; Carayannis, Campbell, & Rehman, 2016). Mode 3 encourages the formation of "creative knowledge environments" (Hemlin, Allwood, & Martin, 2004). "Mode 3 universities," Mode 3 higher education institutions and systems, are prepared to perform "basic research in the context of application" (Campbell & Carayannis, 2013b, p. 34). This has furthermore qualities of nonlinear innovation. Governance of higher education and governance in higher education must also be sensitive, whether a higher education institution operates on the basis of Mode 1, Mode 2, or a combination of these in Mode 3. The concept of "epistemic governance" emphasizes that the underlying knowledge paradigms of knowledge production and knowledge application are being addressed by quality assurance and quality enhancement strategies, policies, and measures (Campbell & Carayannis, 2013b, 2013c).

Emphasizing again a more systemic perspective for the Mode 3 knowledge production, a focused conceptual definition may be as follows (Carayannis & Campbell, 2012, p. 49): Mode 3 "… allows and emphasizes the co-existence and co-evolution of different knowledge and innovation paradigms. In fact, a key hypothesis is: *The competitiveness and superiority of a knowledge system or the degree of advanced development of a knowledge system are highly determined by their adaptive capacity to combine and integrate different knowledge and innovation modes via co-evolution, co-specialization and co-opetition knowledge stock and flow dynamics*" (see Carayannis & Campbell, 2009; on "co-opetition," see Brandenburger & Nalebuff, 1997). Analogies are being drawn and a coevolution is being suggested between diversity and heterogeneity in advanced knowledge society and knowledge economy, and political pluralism in democracy (knowledge democracy), and the quality of a democracy. The "democracy of knowledge" refers to this overlapping relationship. As it is being asserted, "The *Democracy of Knowledge*, as a concept and metaphor, highlights and underscores parallel processes between political pluralism in advanced democracy, and knowledge and innovation heterogeneity and diversity in advanced economy and society. Here, we may observe a hybrid overlapping between the *knowledge economy, knowledge society and knowledge democracy*" (Carayannis & Campbell, 2012, p. 55). The "democracy of knowledge," therefore, is further reaching then the earlier idea of the "Republic of Science" (Michael Polanyi, 1962). This is because there can be a republic that is not democratic, but there cannot be a democracy that is not a democracy (to put here forward a statement in metaphorical terms).

Democracy may be defined as a system that is based on the following principles: freedom, equality, control, and sustainable development (Campbell, Carayannis, & Rehman, 2015). We postulated a coevolution between political systems and innovation systems. Therefore, in this understanding, innovation systems in democracies will differ from innovation systems in nondemocracies. Is there even an expectation of a certain *coevolution between knowledge economy and knowledge democracy*, this ultimately means that certain higher levels of innovation and innovation system are not possible without a context of a democracy (Carayannis & Campbell, 2014). Advanced knowledge economies and knowledge societies require knowledge and innovation pluralism, and this meets with political pluralism in advanced democracies.

The main focus of the Triple Helix innovation model concentrates on university-industry-government relations (Etzkowitz & Leydesdorff, 2000). In that respect, Triple Helix represents a basic model or a core model for knowledge production and innovation application. The models of the Quadruple Helix and Quintuple Helix innovation systems are designed to comprehend already and to refer to an extended complexity in knowledge production and knowledge application (innovation); thus, the analytical architecture of these models is more broadly conceptualized. To use metaphoric terms, the Quadruple Helix embeds and contextualizes the Triple Helix, while the Quintuple Helix embeds and contextualizes the Quadruple Helix (and Triple Helix). The Quadruple Helix adds as a fourth helix the "media-based and culture-based public," the "civil society," and "arts, artistic research, and arts-based innovation" (Campbell, 2018; Carayannis & Campbell, 2009, 2012, p. 14; Carayannis & Campbell, 2018; Carayannis & Pirzadeh, 2014; Campbell & Carayannis, 2016b; see also: Bast, Carayannis, & Campbell, 2015; Danilda, Lindberg, & Torstensson, 2009; De Oliveira Monteiro & Carayannis, 2017; Eigelsreiter, 2017; Hemlin et al., 2004; Mitterlehner, 2014). *The Quadruple Helix also could be emphasized as the perspective that specifically brings in the "dimension of democracy" or the "context of democracy" for knowledge, knowledge production, and innovation.* The Quintuple Helix Innovation Model even is more comprehensive in its analytical and explanatory stretch and approach, adding furthermore the fifth helix (and perspective) of the "natural environments of society" (Carayannis & Campbell, 2010, p. 62) (see Figs. 5.1 and 5.2).

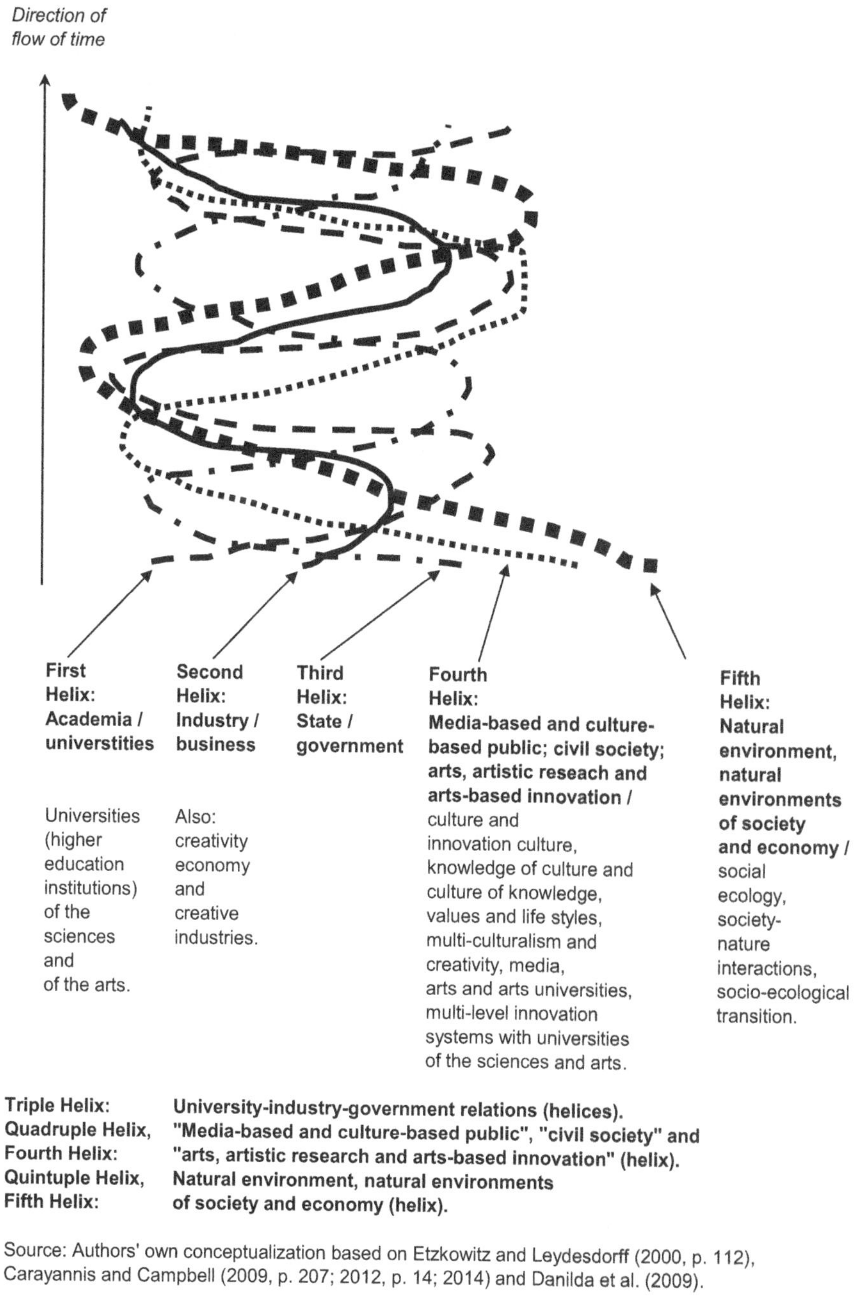

Fig. 5.1 The Quadruple and Quintuple Helix innovation systems. Source: Authors' own conceptualization based on Etzkowitz and Leydesdorff (2000, p. 112), Carayannis and Campbell (2009, p. 207, 2012, p. 14, 2014) and Danilda et al. (2009)

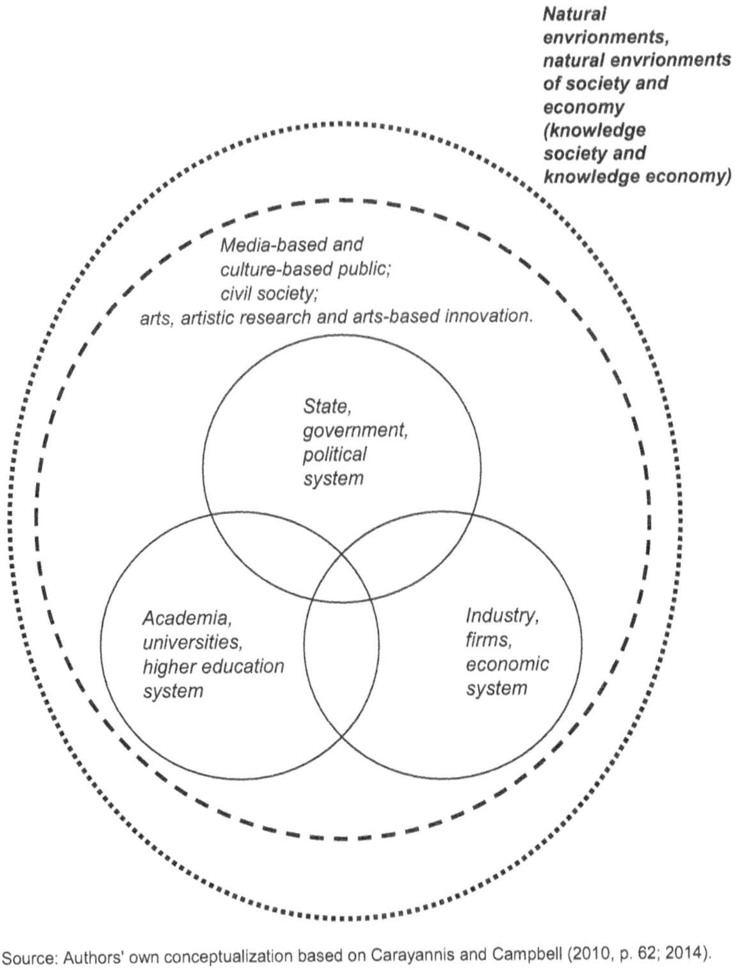

Fig. 5.2 The Quintuple Helix (five-helix model) innovation system more advanced. Source: Authors' own conceptualization based on Carayannis and Campbell (2010, p. 62, 2014)

The introduction of the arts has here two implications: (1) the arts act as a source of creativity, which qualifies as a further necessary input to advance innovation and (2) the different disciplines of the arts extend the established disciplines in the sciences, social sciences, and humanities and by this promoting an extended understanding and new and innovative format of interdisciplinarity but also transdisciplinarity.

The Triple Helix is explicit in acknowledging the importance of higher education for innovation. However, it could be argued that the Triple Helix sees knowledge production and innovation in relation to economy; thus the Triple Helix models first of all (primarily) the economy and economic activity. In that sense, the Triple Helix frames the knowledge economy. The Quadruple Helix brings in the additional perspective of society (knowledge society) and of democracy (knowledge democracy). The Quadruple Helix Innovation System understanding emphasizes that sustainable development of and in economy (knowledge economy) requires that there is a coevolution of knowledge economy and knowledge society and knowledge democracy. The Quadruple Helix even encourages *the perspectives of knowledge society and of knowledge democracy* for supporting, promoting, and advancing knowledge production (research) and knowledge application (innovation). Furthermore, the Quadruple Helix is also explicit that not only universities (higher education institutions) of the sciences but also universities (higher education institutions) of the arts should be regarded as decisive and determining institutions for advancing next-stage innovation systems: the interdisciplinary and transdisciplinary connecting of sciences and arts creates crucial and creative combinations for promoting and supporting innovation. Here, in fact, lies one of the keys for future success. The concept and term of "social ecology" refer to "society-nature interactions" between "human society" and the "material world" (see, e.g., Fischer-Kowalski & Haberl, 2007). The European Commission (2009) identified the necessary socio-ecological transition of economy and society not only as one of the great next-phase challenges but also as an opportunity, for the further progress and advancement of knowledge economy and knowledge society. The Quintuple Helix refers to this socio-ecological transition of society, economy, and democracy, and the Quintuple Helix innovation system is therefore ecologically sensitive. Quintuple Helix bases its understanding of knowledge production (research) and knowledge application (innovation) on social ecology (see Fig. 5.3). Environmental issues (such as global warming) represent issues of concern and of survival for humanity and human civilization. But the Quintuple Helix translates environmental and ecological issues of concern also in potential opportunities, by identifying them as possible drivers for future knowledge production and innovation (Carayannis et al., 2012). This, finally, defines also opportunities for the knowledge economy. "The Quintuple Helix supports here the formation of a win-win situation between ecology, knowledge and innovation, creating synergies between economy, society and democracy" (Carayannis et al., 2012, p. 1).

Summary of the Quadruple and Quintuple Helix Innovation Systems

The terms and concepts of Mode 3 knowledge production and Quadruple Helix innovation systems were first introduced to international academic debate by Carayannis & Campbell (2006, 2009) and were later developed further (Carayannis & Campbell, 2012). The same applies to the Quintuple Helix (Carayannis & Campbell, 2010). From the beginning, the "media-based and culture-based public" as well as universities and other higher education institutions of the arts were being regarded as crucial attributes and components of the Quadruple and Quintuple Helix

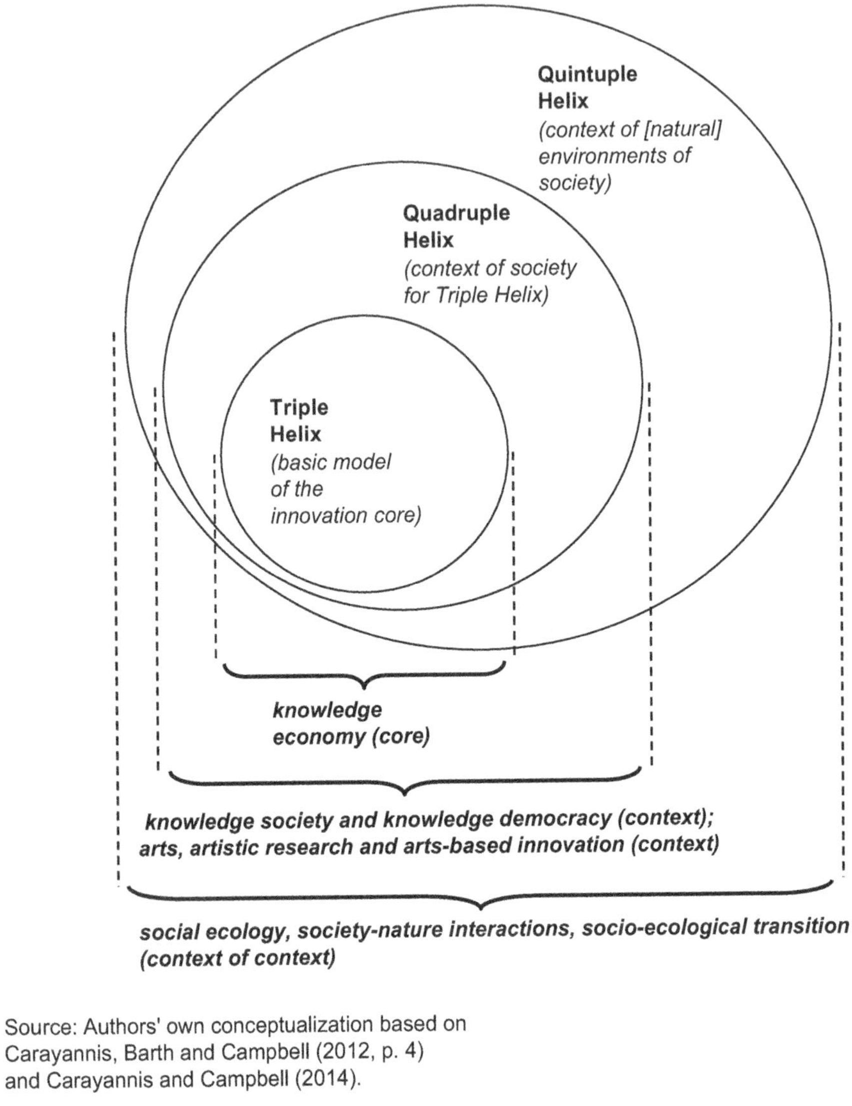

Fig. 5.3 The Quadruple and Quintuple Helix innovation systems in relation to society, economy, democracy, and social ecology. Source: Authors' own conceptualization based on Carayannis et al. (2012, p. 4) and Carayannis and Campbell (2014)

innovation systems, *implying that arts are essential for the progress and evolution of innovation systems* (see again Figs. 5.1 and 5.2). *In our analysis here, we developed more specifically the Quadruple and Quintuple Helix innovation systems in terms and in favor of arts, artistic research, and arts-based innovation. We wanted to demonstrate the full momentum and flexibility of the Quadruple and Quintuple Helix for conceptually addressing and integrating art and arts.*

In the future, what are further challenges for innovation systems? Which issues should be addressed for the design, design evolution, and governance of (and within) innovation systems? More generally speaking, further ramifications of Mode 3 knowledge production in Quadruple Helix and Quintuple Helix innovation systems are (see also Carayannis et al., 2018a, 2018b):

1. *Multilevel innovation systems, the global and the local (GloCal)*: Lundvall was pivotal for introducing the concept of the "national innovation system." Lundvall (1992, p. 1, 3) explicitly acknowledges that national innovation systems are challenged in permanence (but are also extended) by regional as well as global innovation systems. Here, Kuhlmann (2001, p. 960–961) could be paraphrased and the assertion that as long as nation-states and nation-state-based political systems exist, it is plausible to use the concept of the national innovation system. More comprehensive in its analytical architecture than the national innovation system is the concept of the "multilevel innovation system" (Carayannis & Campbell, 2012, p. 32–35). In a spatial understanding, multilevel innovation systems not only compare the national with the sub-national (regional, local) but also with the transnational and global levels (see, e.g., Kaiser & Prange, 2004; furthermore, see Pfeffer, 2012, and Merz & Sormani, 2016). However, it is also important to extend multilevel innovation systems to the challenges and potential benefits and opportunities of a nonspatial meaning, understanding, and "mapping": "Therefore, multi-level systems of knowledge as well as multi-level systems of innovation are based on spatial and non-spatial axes. A further advantage of this multi-level systems architecture is that it results in a more accurate and closer-to-reality description of processes of globalization and *gloCalization*" (Carayannis & Campbell, 2012, p. 35).
2. *Linear and nonlinear innovation*: Knowledge application and innovation are being challenged and driven out of an interest of combining and integrating linear and nonlinear innovation. Key to here are diversity, heterogeneity, and pluralism of different knowledge and innovation modes and their linking together via an architecture of coevolving networks. Firms, universities, and other organizations can engage (at the same) in varying and multiple technology life cycles at different levels of maturity. Another way, how to think nonlinear innovation, is being suggested by the concept of cross-employment (Campbell, 2011, 2013). As a form and type of multi-employment, cross-employment emphasizes that the same individual person may be employed by two (or more) organizations at the same time, where one organization could be located closer to knowledge production and the other to knowledge application (innovation): should those organizations also be rooted in different sectors, then cross-employment acts also as a trans-sectoral networking (Campbell & Carayannis, 2013b, p. 65, 68). *Cross-employment can furthermore bridge different sectors and disciplines in the sciences with different disciplines in the arts.* What results is a "Mode 3 Innovation Ecosystem": "This parallel as well as sequentially time-lagged unfolding of technology life cycles also expresses characteristics of Mode 2 and of nonlinear innovation, because organizations (firms and universities) often must develop

strategies of simultaneously cross-linking different technology life cycles. Universities and firms (commercial and academic firms) must balance the non-triviality of a fluid pluralism of technology life cycles" (Carayannis & Campbell, 2012, p. 37; see furthermore Dubina, Carayannis, & Campbell, 2012). The "academic firm" (Campbell & Carayannis, 2016b) may also be compared with attributes of the so-called network firm (Laperche & Uzunidis, 2018). The relationship between networks, "cooperation and competition" ("co-opetition"), represents a challenge and sensitive issue and allows for different creative answers in organizational representation and manifestation.

3. *Twenty-first century fractal research, education, and innovation ecosystem (FREIE)*: Here, the understanding of FREIE is: "This is *a multilayered, multimodal, multinodal, and multilateral system*, encompassing mutually complementary and reinforcing innovation networks and knowledge clusters consisting of human and intellectual capital, shaped by social capital and underpinned by financial capital" (Carayannis & Campbell, 2012, p. 3).
4. *Linear and nonlinear innovation, and the causality of "if-then" and of "if-if" relations*: The hybrid overlapping of linear innovation and of nonlinear innovation displays also possible ramifications and draws associations to models of causality and their remodeling. "We can speculate, whether this parallel integration of linearity and nonlinearity not also encourages a new approach of paralleling in our theorizing and viewing of causality: *in epistemic (epistemological) terms, the so-called if-then relationships could be complemented by (a thinking in) 'if-if' relations*" (Carayannis & Campbell, 2012, p. 24; see also Campbell, 2009, p. 123).

The Quadruple Helix regards itself to be "human-centered" oriented. While for the Triple Helix model the existence of a democracy is not (per se) necessary for knowledge production and innovation, the Quadruple Helix is here more explicit. With the way how the Quadruple Helix is being engineered, designed, and "architected" from that, it is clear that there cannot be a Quadruple Helix Innovation System without democracy or a democratic context. The following attributes and components define the fourth helix in the Quadruple Helix: "media-based and culture-based public," "civil society," and "arts, artistic research, and arts-based innovation." By this the fourth helix in the Quadruple Helix represents the perspective of the "dimension of democracy" or the "context of democracy" for knowledge, knowledge production, and innovation. This is particularly true when democracy is being understood to transcend the narrow understanding of being primarily based *on* or being primarily rooted *in* government institutions (within Triple Helix). Civil society, culture-based public, quality of democracy, and sustainable development convincingly demonstrate what the rationales and requirements are for conceptualizing democracy broader (Campbell & Carayannis, 2013a). To turn this line of thinking, autocracies are not interested to allow the development of a free and mature civil society. On the contrary, autocracies want to control and suppress the rise of an independent civil society. *Political pluralism in a democracy coevolves with the pluralism, diversity, and heterogeneity of knowledge, knowledge production, and innovation* ("democracy of knowledge"; see Carayannis & Campbell, 2009, 2012, p. 55). *We postulate here a congruence of structures and processes in democracy and in innovation systems.* The Quintuple Helix extends the Quadruple Helix by aspects of the "natural environments of society and economy," "social ecology," and the "socio-ecological transition." Also, this environmental context of society can be better addressed in a democracy than in a nondemocracy. *The current world appears to be challenged by a race between developing democracies versus emerging autocracies over knowledge production and innovation.*

Cyber development can be defined as a development in terms of a sustainable development of knowledge economy, knowledge society, and knowledge democracy that is knowledge-based and knowledge-driven and where innovation is playing a crucial role. In this understanding, the Quadruple and Quintuple Helix Innovation System and systems provide a model and conceptual framework for theory and practice, strategy, and policy for progress and advancement exactly in knowledge economy, knowledge society, and knowledge democracy. This introduces new perspectives for a new type of governance and a new set of policies for problem-solving and further evolution.

Chapter 6
Conclusion: Smart Quintuple Helix Innovation Systems

The Lancet Commission (2017) has released a critical report on the ecological status of the world, with the following core assessment:

> Pollution is the largest environmental cause of disease and premature death in the world today. Diseases caused by pollution were responsible for an estimated 9 million premature deaths in 2015—16% of all deaths worldwide—three times more deaths than from AIDS, tuberculosis, and malaria combined and 15 times more than from all wars and other forms of violence. (Lancet Commission, 2017, p. 1)

In the report "State of the Climate in 2017," published by the American Meteorological Society in August 2018, it is being stated:

> In 2017, the dominant greenhouse gases released into Earth's atmosphere—carbon dioxide, methane, and nitrous oxide— reached new record highs. The annual global average carbon dioxide concentration at Earth's surface for 2017 was 405.0 ± 0.1 ppm, 2.2 ppm greater than for 2016 and the highest in the modern atmospheric measurement record and in ice core records dating back as far as 800,000 years. The global growth rate of CO_2 has nearly quadrupled since the early 1960s.... In the Arctic, the 2017 land surface temperature was 1.6 °C above the 1981–2010 average, the second highest since the record began in 1900, behind only 2016. The five highest annual Arctic temperatures have all occurred since 2007. (Blunden, Arndt, & Hartfield, 2018, p. Sxvi)

The ecological indicators show that there is more of a global warming, where the released greenhouse gases are increasing and where also the temperatures are continuously rising further. This is even feeding speculations, whether the world may be entering a so-called "Hothouse Earth" scenario:

> We explore the risk that self-reinforcing feedbacks could push the Earth System toward a planetary threshold that, if crossed, could prevent stabilization of the climate at intermediate temperature rises and cause continued warming on a 'Hothouse Earth' pathway even as human emissions are reduced. Crossing the threshold would lead to a much higher global average temperature than any interglacial in the past 1.2 million years and to sea levels significantly higher than at any time in the Holocene.... Collective human action is required to steer the Earth System away from a potential threshold and stabilize it in a habitable interglacial-like state. Such action entails stewardship of the entire Earth System—biosphere, climate, and societies—and could include decarbonization of the global economy,

E. G. Carayannis, D. F. J. Campbell, *Smart Quintuple Helix Innovation Systems*, SpringerBriefs in Business, https://doi.org/10.1007/978-3-030-01517-6_6

enhancement of biosphere carbon sinks, behavioral changes, technological innovations, new governance arrangements, and transformed social values. (Steffen et al., 2018, p. 1)

In summary, it should be clear that all systems (subsystems, sectors, and subsectors) in a Quintuple Helix and a Smart Quintuple Helix Innovation System are performing a pivotal function, influencing each other. If more sustainable development is being considered (and demanded) on a national level, as a result of "global warming" (Carayannis, 2011), and if, for instance, more targeted investments in a specific Helix of the Quintuple Helix start flowing, then there will be a positive impact on all other subsystems and on the society as a whole. The Quintuple Helix model demonstrates that an investment in knowledge and a promotion of knowledge production bring into play new and crucial impulses for innovation, know-how, and the advancement of society. By initiating small steps toward sustainability, long-term and leading knowledge societies can emerge, which will live in balance with nature, and ultimately, perhaps, lead to a "green economic wonder."

To conclude, the Quintuple Helix model makes it clear that the implementation of thought and action in sustainability will have a positive impact on society as a whole. The new quality management for more sustainability lies therefore in the creation of new knowledge, know-how, and innovation in balance with nature (see Carayannis & Campbell, 2010, pp. 58–62). One chief objective of the Quintuple Helix is to enhance "value in society" through the resource of knowledge. The discussion about the Quintuple Helix model indicates that striving for the promotion of knowledge as a "knowledge nugget" should be regarded as being essential (see Carayannis & Formica, 2006, p. 152): This means that knowledge is the key to and fore more sustainability and to a new quality of life. Today, knowledge is the "most fundamental resource" (Lundvall, 1992, p. 1). Nevertheless, whether a state (nation-state, beyond nation-state) is leading in different fields in the future will be primarily, if not even solely, decided by its potential to develop new knowledge, know-how, and innovation in balance with nature. However, the improved exchange of knowledge and the striving for knowledge, new know-how, and innovations through the Quintuple Helix model can be or at least offer a solution for the challenges of sustainable development under the aspect of global warming in the twenty-first century.

Mastering and balancing ecological issues and challenges (such as global warming) often are being depicted and presented as a theme of survival for humanity in a global format (e.g., see Steffen et al., 2018). Consequently, the European Commission (2009) can assert the major need for a greater "socio-ecological transition." Social ecology makes the context of the natural environments for society and economy more visible and emphasizes an understanding of interaction and co-development of society and environment (nature). The Quintuple Helix Innovation Model (Carayannis & Campbell, 2010) bridges social ecology with knowledge production and innovation. *Here, the natural environments of society and economy challenge but also encourage and inspire knowledge production and innovation. In the approach of the Quintuple Helix Innovation Model, the natural environments of society are being identified as opportunities for driving further and excelling the sustainable development and coevolution of knowledge economy, knowledge society, and knowledge democracy*. This also has a potential of influencing the way how we perceive and organize entrepreneurship.

Recent empirical evidence suggests further that there is by tendency a decoupling of economic growth from a further increase of energy sector emissions: "Private-sector incentives help drive decoupling of emissions and economic growth. The importance of this trend cannot be understated. This 'decoupling' of energy sector emissions and economic growth should put to rest the argument that combatting climate change requires accepting lower growth or a lower standard of living" (Obama, 2017, p. 1). *There is evidence that there can be and that there actually is economic growth and economic development that is environmentally and ecologically sensitive and that supports (and encourages) environmental protection* (see again the summary and presented overview in Obama, 2017). *This certainly indicates a road further into a promising and progressive future of human civilization (on this planet and beyond), where economic growth, sustainable development, and quality of democracy are coming together, where there is a coevolution of knowledge economy, knowledge society, and knowledge democracy* (Campbell, 2018). Eco-innovation and eco-entrepreneurship within diverse innovation ecosystems serve here as interesting and innovative examples (Carayannis, Barth, & Campbell, 2012).

The Quadruple Helix and Quintuple Helix regard themselves to be "human-centered" oriented. While for the Triple Helix model the existence of a democracy is not (per se) necessary for knowledge production and innovation, the Quadruple Helix is here more explicit. With the way how the Quadruple Helix is being engineered, designed, and "architected," from that it is clear that there cannot be a Quadruple Helix Innovation System without democracy or a democratic context. The following attributes and components define the fourth helix in the Quadruple Helix: "media-based and culture-based public," "civil society," and "arts, artistic research, and arts-based innovation." By this the fourth helix in the Quadruple Helix represents the perspective of the "dimension of democracy" or the "context of democracy" for knowledge, knowledge production, and innovation. This is particularly true when democracy is being understood to transcend the narrow understanding of being primarily based *on* or being primarily rooted *in* government institutions (within Triple Helix). Civil society, culture-based public, quality of democracy, and sustainable development convincingly demonstrate what the rationales and requirements are for conceptualizing democracy broader (Campbell & Carayannis, 2013a).[1] *Political pluralism in a democracy coevolves with the pluralism, diversity, and heterogeneity of knowledge, knowledge production, and innovation* ("Democracy of Knowledge," see Carayannis & Campbell, 2009, 2012, p. 55). *We postulate here a congruence of structures and processes in democracy and in innovation systems*. The Quintuple Helix extends the Quadruple Helix by aspects of the "natural environments of society and economy," "social ecology," and the "socio-ecological transition." Also this environmental context of society can be better addressed in a

[1] To turn this line of thinking, autocracies are not interested to allow the development of a free and mature civil society. On the contrary, autocracies want to control and suppress the rise of an independent civil society.

democracy than in a nondemocracy. *The current world appears to be challenged by a race between developing democracies versus emerging autocracies over knowledge production and innovation* (Carayannis & Campbell, 2014, p 19).

At the beginning of our analysis, we have listed two research questions. The first research question was carrying the core of our whole analysis: *How do knowledge, innovation, and the environment (natural environment) relate to each other?* The second research question we have posed is: *What are key features of Smart Quintuple Helix Innovation Systems?* We will address now this second research question in terms of a draft and in the design making of a short "manifesto," where we are putting forward several propositions to be further discussed.

The theory, concept, and program of the Smart Quintuple Helix Innovation Systems refer to the following features, which can be regarded as hypotheses for strategy, policy-making, and decision-making.

1. Environmental and ecological sensitivity should be regarded as key drivers for knowledge, knowledge creation, and knowledge production and innovation. So how can knowledge and innovation, which are environmentally and ecologically sensitive, be translated into economic success, economic growth, and economic development?
2. Long-term economic development and sustainable development are furthermore based on ecological and environmental sensitivity and finally have the potential to provide for higher or for more stable economic growth rates. This also refers to social ecology.
3. In the long run, there is more (and not less) economic growth in combination with environmentally (ecologically) friendly economic development and sustainable development. This represents a core assumption.
4. Environmental protection must be regarded also as an input and an investment in economic growth and for economic growth.
5. There is no alternative to democracy, not now and not in the future. Democracy is the future. Democracy encourages knowledge and innovation for quality of democracy within the framework of a knowledge democracy.
6. Knowledge economy, knowledge society, and knowledge democracy are based on knowledge and innovation, and (a) they understand knowledge and innovation as key drivers for economic growth and economic development, and (b) they understand environmental and ecological sensitivity as key drivers for knowledge and innovation.
7. *Smart (SmART, smart art) Quintuple Helix Innovation Systems see and appreciate "art" also as a manifestation of knowledge and innovation (in addition to art as a manifestation of aesthetics).*

References

Adams, W. M. (2006). The future of sustainability: Re-thinking environment and development in the twenty-first century. *Report of the IUCN Renowned Thinkers Meeting, January 29-21, 2006*. Madrid, Spain: The World Conservation Union (IUCN). Retrieved from http://cmsdata.iucn.org/downloads/iucn_future_of_sustainability.pdf

Barth, T. D. (2011a). The idea of a green new deal in a Quintuple Helix Model of knowledge, know-how and innovation. *International Journal of Social Ecology and Sustainable Development, 2*(2), 1–14.

Barth, T. D. (2011b). *Freiheit, Gleichheit, Demokratiequalität: Zur Qualitätsmessung in den Top 20 Demokratien des Democracy Rankings [Freedom, equality and the quality of democracy: Measuring quality in the top 20 democracies of the democracy ranking]* (Doctoral thesis, University of Vienna, Vienna, Austria).

Barth, T. D. (2011c). Die 20 besten Demokratien der Welt. *Freiheit – Gleichheit – Demokratiequalität auf einen Blick* (1. Auflage). Norderstedt, Germany: Demand Verlag.

Bast, G. (2013). Preparing a "creative revolution" – arts and universities of the arts in the creative knowledge economy. In E. G. Carayannis (Editor-in-Chief), I. N. Dubina, N. Seel, D. F. J. Campbell, & D. Uzunidis (Associate Editors) (Eds.), *Encyclopedia of creativity, invention, innovation and entrepreneurship* (pp. 1471–1476). New York, NY: Springer. https://doi.org/10.1007/978-1-4614-3858-8_442.

Bast, G., Carayannis, E. G., & Campbell, D. F. J. (Eds.). (2015). *Arts, research, innovation and society*. New York, NY: Springer. Retrieved from http://www.springer.com/business+%26+management/technology+management/book/978-3-319-09908-8

Baum, S., & Payea, K. (2005). *Education pays 2005 update*. Washington, D.C.: College Board. Retrieved from http://www.collegeboard.com/prod_downloads/press/cost05/education_pays_05.pdf

Baum, S., Payea, K., & Steele, P. (2006). *Education pays 2006 second update*. Washington, DC: College Board. Retrieved from http://www.collegeboard.com/prod_downloads/press/cost06/education_pays_06.pdf

Bhaskar, R. (2010). Context of interdisciplinarity: Interdisciplinarity and climate change. In R. Bhasakar, C. Frank, K. G. Høyer, P. Næss, & J. Parker (Eds.), *Interdisciplinarity and climate change: Transforming knowledge and practice for our global future* (pp. 1–24). New York, NY: Routledge.

Blunden, J., Arndt, D. S., & Hartfield, G. (Eds.). (2018). State of the climate in 2017. *Bulletin of the American Meteorological Society 99*(8), Si-S332. https://doi.org/10.1175/2018BAMSStateoftheClimate.1. Retrieved from https://www.ametsoc.org/ams/index.cfm/publications/bulletin-of-the-american-meteorological-society-bams/state-of-the-climate/

E. G. Carayannis, D. F. J. Campbell, *Smart Quintuple Helix Innovation Systems*, SpringerBriefs in Business, https://doi.org/10.1007/978-3-030-01517-6

Bomberg, E., & Stubb, A. (2003). Introduction. In E. Bomberg & A. Stubb (Eds.), *The European Union: How does it work?* (pp. 3–18). Oxford, UK: Oxford University Press.

Boulding, K. E. (1985). *The world as a total system*. Beverly Hills, CA: Sage.

Brandenburger, A. M., & Nalebuff, B. J. (1997). *Co-opetition*. New York, NY: Doubleday.

Bush, V. (1945). *Science: The endless frontier*. Washington, DC: United States Government Printing Office. Retrieved from http://www.nsf.gov/od/lpa/nsf50/vbush1945.htm#transmittal

Campbell, D. F. J. (2008). *The basic concept for the democracy ranking of the quality of democracy*. Vienna, Austria: Democracy Ranking. Retrieved from http://www.democracyranking.org/downloads/basic_concept_democracy_ranking_2008_A4.pdf

Campbell, D. F. J. (2009). Externe Umwelten. Außensichten auf das iff. In M. Arnold (Ed.). Iff. *Interdisziplinäre Wissenschaft im Wandel* (pp. 101–134). Vienna, Austria: LIT.

Campbell, D. F. J. (2011). Wissenschaftliche „Parallelkarrieren" als Chance. Wenn Wissenschaft immer öfter zur Halbtagsbeschäftigung wird, könnte eine Lösung im „Cross-Employment" liegen. Guest Commentary for DIE PRESSE (February 2, 2011). Retrieved from http://diepresse.com/home/bildung/meinung/635781/Wissenschaftliche-Parallelkarrieren-als-Chance?direct=635777&_vl_backlink=/home/bildung/index.do&selChannel=500

Campbell, D. F. J. (2013). Cross-employment. In E. G. Carayannis (Editor-in-Chief), I. N. Dubina, N. Seel, D. F. J. Campbell, & D. Uzunidis (Associate Editors) (Eds.). *Encyclopedia of creativity, invention, innovation and entrepreneurship* (pp. 503–508). New York, NY: Springer. https://doi.org/10.1007/978-1-4614-3858-8_254.

Campbell, D. F. J. (2018). *Global quality of democracy as innovation enabler. Measuring democracy for success*. New York, NY: Palgrave Macmillan. Retrieved from https://www.palgrave.com/de/book/9783319725284

Campbell, D. F. J., & Barth, T. D. (2009). Wie können Demokratie und Demokratiequalität gemessen werden? Modelle, Demokratie-Indices und Länderbeispiele im globalen Vergleich. *SWS-Rundschau 49*(2), 208–233. Retrieved from http://www.uni-klu.ac.at/wiho/downloads/campbell_u._barth-demokratiemessung-sws_rundschau-heft_2009_02-FINAL.pdf

Campbell, D. F. J., & Carayannis, E. G. (2012). Lineare und nicht-lineare Knowledge Production: Innovative Herausforderungen für das Hochschulsystem. *Zeitschrift für Hochschulentwicklung 7*(2), 64–72. Retrieved from http://www.zfhe.at/index.php/zfhe/article/view/448

Campbell, D. F. J., & Carayannis, E. G. (2013a). Quality of democracy and innovation. In E. G. Carayannis (Editor-in-Chief), I. N. Dubina, N. Seel, D. F. J. Campbell, & D. Uzunidis (Associate Editors) (Eds.), *Encyclopedia of creativity, invention, innovation and entrepreneurship* (pp. 1527–1534). New York, NY: Springer. https://doi.org/10.1007/978-1-4614-3858-8_509.

Campbell, D. F. J., & Carayannis, E. G. (2013b). *Epistemic governance in higher education. Quality enhancement of universities for development*. SpringerBriefs in Business. New York, NY: Springer. Retrieved from http://www.springer.com/business+%26+management/organization/book/978-1-4614-4417-6

Campbell, D. F. J., & Carayannis, E. G. (2013c). Epistemic governance and epistemic innovation policy. In E. G. Carayannis (Editor-in-Chief), I. N. Dubina, N. Seel, D. F. J. Campbell, & D. Uzunidis (Associate Editors) (Eds.), *Encyclopedia of creativity, invention, innovation and entrepreneurship* (pp. 697–702). New York, NY: Springer. https://doi.org/10.1007/978-1-4614-3858-8_271

Campbell, D. F. J., & Carayannis, E. G. (2015). Explaining and comparing quality of democracy in quadruple helix structures: The quality of democracy in the United States and in Austria, challenges and opportunities for development. In E. G. Carayannis, D. F. J. Campbell, & M. P. Efthymiopoulos (Eds.), *Cyber-development, cyber-democracy and cyber-defense. Challenges, opportunities and implications for theory, policy and practice*. New York, NY: Springer. Retrieved from http://www.springer.com/social+sciences/political+science/book/978-1-4939-1027-4

Campbell, D. F. J., & Carayannis, E. G. (2016a). Epistemic governance and epistemic innovation policy. *Technology, Innovation and Education 2*, 1–15. https://doi.org/10.1186/s40660-016-0008-2. Retrieved from http://technology-innovation-education.springeropen.com/articles/10.1186/s40660-016-0008-2

Campbell, D. F. J., & Carayannis, E. G. (2016b). The academic firm: A new design and redesign proposition for entrepreneurship in innovation-driven knowledge economy. *Journal of Innovation and Entrepreneurship* *5*(12), 1–10. https://doi.org/10.1186/s13731-016-0040-1. Retrieved from http://innovation-entrepreneurship.springeropen.com/articles/10.1186/s13731-016-0040-1

Campbell, D. F. J., & Pölzlbauer, G. (2009). *The democracy ranking 2008/2009 of the quality of democracy: Method and ranking outcome*. Vienna, Austria: Democracy Ranking. Retrieved from http://www.democracyranking.org/downloads/method_ranking_outcome_2008-09_A4.pdf

Campbell, D. F. J., & Miklós Sükösd (Eds.) (2002). *Feasibility study for a quality ranking of democracies*. Vienna, Austria: Global Democracy Award. Retrieved from http://www.democracyranking.org/downloads/feasibility_study-a4-e-01.pdf

Campbell, D. F. J. (2001). Politische Steuerung über öffentliche Förderung universitärer Forschung? Systemtheoretische Überlegungen zu Forschungs- und Technologiepolitik. [Political Steering through Public Support of University Research? Systems-Theoretical Considerations about Research and Technology Policy]. *Österreichische Zeitschrift für Politikwissenschaft (Austrian Journal of Political Science), 30*(4), 425–438.

Campbell, D. F. J. (2006). The university/business research networks in science and technology: Knowledge production trends in the United States, European Union and Japan. In E. G. Carayannis & D. F. J. Campbell (Eds.), *Knowledge creation, diffusion, and use in innovation networks and knowledge clusters. A comparative systems approach across the United States, Europe and Asia* (pp. 67–100). Westport, CT: Praeger.

Campbell, D. F. J., Carayannis, E. G., & Rehman, S. S. (2015). Quadruple helix structures of quality of democracy in innovation systems: The USA, OECD countries, and EU member countries in global comparison. *Journal of the Knowledge Economy, 6*(3), 467–493. https://doi.org/10.1007/s13132-015-0246-7

Campbell, D. F. J., & Güttel, W. H. (2005). Knowledge production of firms: Research networks and the "scientification" of business R&D. *International Journal of Technology Management, 31*(1-2), 152–175.

Campbell, G. S. (2013b). Speaking pictures: Innovation in fine arts. In E. G. Carayannis (Editor-in-Chief), I. N. Dubina, N. Seel, D. F. J. Campbell, & D. Uzunidis (Associate Editors) (Eds.), *Encyclopedia of creativity, invention, innovation and entrepreneurship* (pp. 1716–1722). New York, NY: Springer. https://doi.org/10.1007/978-1-4614-3858-8_484.

Carayannis, E. G. (Ed.) (2011). *Planet earth 2011. Global warming challenges and opportunities for policy and practice*. London, United Kingdom: InTech. Retrieved from http://www.intechopen.com/books/show/title/planet-earth-2011-global-warming-challenges-and-opportunities-for-policy-and-practice

Carayannis, E. G., & Alexander, J. M. (2006). *Global and local knowledge. Glocal transatlantic public-private partnerships for research and technological development*. Basingstoke, England: Palgrave MacMillan.

Carayannis, E. G., Barth, T. D., & Campbell, D. F. J. (2012). The Quintuple Helix innovation model: Global warming as a challenge and driver for innovation. *Journal of Innovation and Entrepreneurship* *1*(1), 1–12. Retrieved from http://www.innovation-entrepreneurship.com/content/pdf/2192-5372-1-2.pdf

Carayannis, E. G., & Campbell, D. F. J. (2006). "Mode 3": Meaning and implications from a knowledge systems perspective. In E. G. Carayannis & D. F. J. Campbell (Eds.), Knowledge *creation, diffusion, and use in innovation networks and knowledge clusters. A comparative systems approach across the United States, Europe and Asia*. (pp. 1–25). Westport, CT: Praeger.

Carayannis, E. G., & D. F. J. Campbell (2009). "Mode 3" and "Quadruple Helix": Toward a 21st century fractal innovation ecosystem. *International Journal of Technology Management* *46*(3-4), 201–234. Retrieved from http://www.inderscience.com/browse/index.php?journalID=27&year=2009&vol=46&issue=3/4 and http://www.inderscience.com/search/index.php?action=record&rec_id=23374&prevQuery=&ps=10&m=or

Carayannis, E. G., & Campbell, D. F. J. (2010). Triple Helix, Quadruple Helix and Quintuple Helix and how do knowledge, innovation and the environment relate to each other? a proposed framework for a trans-disciplinary analysis of sustainable development and social ecology. *International Journal of Social Ecology and Sustainable Development 1*(1), 41–69. Retrieved from https://www.igi-global.com/article/triple-helix-quadruple-helix-quintuple/41959

Carayannis, E. G., & Campbell, D. F. J. (2012). *Mode 3 knowledge production in quadruple helix innovation systems. 21st-century democracy, innovation, and entrepreneurship for development.* SpringerBriefs in Business. New York, NY: Springer. Retrieved from http://www.springer.com/business+%26+management/book/978-1-4614-2061-3

Carayannis, E. G., & Campbell, D. F. J. (2014). Developed democracies versus emerging autocracies: Arts, democracy, and innovation in quadruple helix innovation systems. *Journal of Innovation and Entrepreneurship 3*(12), 1–23. Retrieved from http://www.innovation-entrepreneurship.com/content/3/1/12

Carayannis, E. G., Campbell, D. F. J., Rehman, S. S. (2016). Mode 3 knowledge production: Systems and systems theory, clusters and networks. *Journal of Innovation and Entrepreneurship 5*(17), 1–24. https://doi.org/10.1186/s13731-016-0045-9. Retrieved from http://innovation-entrepreneurship.springeropen.com/articles/10.1186/s13731-016-0045-9

Carayannis, E. G., David F. J, & Campbell, D. F. J. (2011). Open innovation diplomacy and a 21st century fractal research, education and innovation (FREIE) ecosystem: Building on the quadruple and Quintuple Helix innovation concepts and the "Mode 3" knowledge production system. *Journal of the Knowledge Economy 2*(3), 327–372. Retrieved from http://www.springerlink.com/content/d1lr223321305579/

Carayannis, E. G., & Formica, P. (2006). Intellectual venture capitalists: An emerging breed of knowledge entrepreneurs – viewpoint. *Industry and Higher Education, 20*(3), 151–156.

Carayannis, E. G., Grigoroudis, E., Campbell, D. F. J., Meissner, D., & Stamati, D. (2018a). 'Mode 3' universities and academic firms: Thinking beyond the box trans-disciplinarity and nonlinear innovation dynamics within coopetitive entrepreneurial ecosystems. *International Journal of Technology Management 77*(1-3), 145–185. Retrieved from https://www.inderscienceonline.com/doi/abs/10.1504/IJTM.2018.091714 und https://www.inderscienceonline.com/toc/ijtm/77/1-3

Carayannis, E. G., & Pirzadeh, A. (2014). *The knowledge of culture and the culture of knowledge. implications for theory, policy and practice.* Basingstoke, England: Palgrave Macmillan. Retrieved from http://www.amazon.de/The-Knowledge-Culture-Implications-Practice/dp/1403942439/ref=sr_1_1?ie=UTF8&qid=1403080044&sr=8-1&keywords=carayannis+knowledge+of+culture

Carayannis, E. G., & Von Zedtwitz, M. (2005). Architecting gloCal (global – local), real-virtual incubator networks (G-RVINs) as catalysts and accelerators of entrepreneurship in transitioning and developing economies. *Technovation, 25*, 95–110.

Carayannis, E. G., & Campbell, D. F. J. (2007). A "Mode 3" systems approach for knowledge creation, diffusion, and use: Towards a twenty-first-century fractal innovation ecosystem. In E. G. Carayannis & C. Ziemnowicz (Eds.), *Rediscovering Schumpeter. Creative destruction evolving into "Mode 3"* (pp. 71–111). Basingstoke, England: Palgrave Macmillan.

Carayannis, E. G., & Campbell, D. F. J. (2018). Quadruple and Quintuple Helix innovation systems and mode 3 knowledge 3 production. In E. G. Carayannis, D. F. J. Campbell, & M. P. Efthymiopoulos (Eds.), *Handbook of cyber-development, cyber-democracy, and cyber-defense* (pp. 9–27). New York, NY: Springer. https://doi.org/10.1007/978-3-319-06091-0

Carayannis, E. G., Grigoroudis, E., Campbell, D. F. J., Meissner, D., & Stamati, D. (2018b). The ecosystem as helix: An exploratory theory-building study of regional co-opetitive entrepreneurial ecosystems as Quadruple/Quintuple Helix Innovation Models. *R&D Management, 48*(1), 148–162. https://doi.org/10.1111/radm.12300

Carayannis, E. G., & Kaloudis, A. (2010). A time for action and a time to lead: Democratic capitalism and a new "new deal" for the US and the world in the twenty-first century. *Journal of the Knowledge Economy, 1*(1), 4–17. https://doi.org/10.1007/s13132-009-0002-y

Carayannis, E. G., & Ziemnowicz, C. (Eds.). (2007). *Re-discovering Schumpeter: Creative destruction evolving into "Mode 3"*. London, UK: Palgrave Macmillan.

Carmines, E. G., & Stimson, J. A. (1980). The two faces of issue voting. *The American Political Science Review, 74*(1), 78–91.

Cullell, J. V. (2004). Democracy and the quality of democracy. Empirical findings and methodological and theoretical issues drawn from the citizen audit of the quality of democracy in Costa Rica. In G. O'Donnell, J. V. Cullell, & O. M. Iazzetta (Eds.), *The quality of democracy. Theory and applications* (pp. 93–162). Notre Dame, IN: University of Notre Dame Press.

Dahl, R. A. (1971). *Polyarchy. Participation and opposition.* New Haven, CT: Yale University Press.

Danilda, I., Lindberg, M., Torstensson, B.-M. (2009). *Women resource centres. A quattro helix innovation system on the European Agenda.* Paper. Retrieved from http://www.hss09.se/own_documents/Papers/3-11%20-%20Danilda%20Lindberg%20&%20Torstensson%20-%20paper.pdf

De Oliveira Monteiro, S. P., & Carayannis, E. G. (Eds.). (2017). *The quadruple innovation helix nexus. A smart growth model, qualitative empirical validation and operationalization for OECD countries.* New York, NY: Palgrave Macmillan.

Diamond, L., & Morlino, L. (2004). The quality of democracy. An overview. *Journal of Democracy, 15*(4), 20–31.

Dubina, I. N. (2009). *Управление творчеством персонала в условиях инновационной экономики [Creativity Management in the Innovation Economy].* Moscow, Russia: Academia.

Dubina, I. N. (2005). Managing creativity: Theoretical approaches to employees' creativity development and regulation. *International Journal of Management Concepts and Philosophy, 1*(4), 334–349.

Dubina, I. N. (2007). Optimally managing creativity in organisations. In E. G. Carayannis (Ed.), *Managing creative and innovative people: The art, science and craft of fostering creativity, triggering invention and catalyzing innovation* (pp. 143–170). Westport, CT: Praeger.

Dubina, I. N., Carayannis, E. G., & Campbell, D. F. J. (2012). Creativity economy and a crisis of the economy? Coevolution of knowledge, innovation, and creativity, and of the knowledge economy and knowledge society. *Journal of the Knowledge Economy, 3*(1), 1–24. https://doi.org/10.1007/s13132-011-0042-y

Eigelsreiter, B. (2017). Consumerization of IT, cyber-democracy and cyber-crime: The inherent challenges and opportunities of different ends of a continuum. In E. G. Carayannis, D. F. J. Campbell, & M. P. Efthymiopoulos (Eds.), *Handbook of cyber-development, cyber-democracy, and cyber-defense*. New York, NY: Springer. https://doi.org/10.1007/978-3-319-06091-0

Esty, D. C., Levy, M., Srebotnjak, T., de Sherbinin, A. (2005). *2005 Environmental sustainability index: Benchmarking national environmental stewardship.* New Haven, CT: Yale Center for Environmental Law & Policy. Retrieved from http://www.yale.edu/esi/ESI2005_Main_Report.pdf

Esty, D. C., Levy, M. A., Kim, C. H., de Sherbinin, A., Srebotnjak, T., & Mara, V. (2008). *2008 Environmental performance index.* New Haven, CT: Yale Center for Environmental Law and Policy. Retrieved from http://www.yale.edu/epi/files/2008EPI_Text.pdf

Etzkowitz, H. (2003). Research groups as "quasi-firms": The invention of the entrepreneurial university. *Research Policy, 32*, 109–121.

Etzkowitz, H., & Leydesdorff, L. (2000). The dynamics of innovation: From National Systems and "Mode 2" to a Triple Helix of university-industry-government relations. *Research Policy, 29*, 109–123.

European Commission. (2009). *The world in 2025. Rising Asia and socio-ecological transition.* Brussels, Belgium: Author. Retrieved from http://ec.europa.eu/research/social-sciences/pdf/the-world-in-2025-report_en.pdf

Fischer-Kowalski, M. (1998). Society's metabolism. The intellectual history of materials flow analysis, part I, 1860-1970. *Journal of Industrial Ecology, 2*(1), 61–78.

Fischer-Kowalski, M., & Haberl, H. (Eds.). (2007). *Socioecological transitions and global change. Trajectories of social metabolism and land use.* Cheltenham, England: Edward Elgar.

Fischer-Kowalski, M., & Hüttler, W. (1999). Society's metabolism. The intellectual history of materials flow analysis, part II, 1970-1998. *Journal of Industrial Ecology, 2*(4), 107–136.

Florida, R. (2004). *The rise of the creative class: And how it's transforming work, leisure, community, and everyday life*. Cambridge, MA: Basic Books.

Freedom House. (2008). *Methodology*. Washington, DC: Author. Retrieved from http://www.freedomhouse.org/template.cfm?page=351&ana_page=341&year=2008

Freedom House. (2009a). *Freedom in the world comparative and historical data. Country ratings and status, FIW (freedom in the world) 1973-2009*. Washington, DC: Author. Retrieved from http://www.freedomhouse.org/template.cfm?page=439

Freedom House. (2009b). *Map of freedom in the world*. Washington, DC: Author. Retrieved from http://www.freedomhouse.org/template.cfm?page=363&year=2009

Friedman, T. L. (2008). *Hot, flat and crowded: Why we need a green revolution – and how we can renew our global future*. London, UK: Penguin Books.

Gibbons, M., Limoges, C., Nowotny, H., Schwartzman, S., Scott, P., & Trow, M. (1994). *The new production of knowledge. The dynamics of science and research in contemporary societies*. London, UK: Sage.

Haberl, H., Fischer-Kowalski, M., Krausmann, F., Martinez-Alier, J., & Winiwarter, V. (2009). A socio-metabolic transition towards sustainability? Challenges for another great transformation. *Sustainable Development, 19*(1), 1.

Haberl, H., Fischer-Kowalski, M., Krausmann, F., Weisz, H., & Winiwarter, V. (2004). Progress towards sustainability? What the conceptual framework of material and energy flow accounting (MEFA) can offer. *Land Use Policy, 21*(3), 199–213.

Harding, S., Phillips, D., & Fogarty, M. (1986). Contrasting values in western Europe. Unity, diversity and change. In *Studies in the contemporary values of modern society*. Basingstoke, England: MacMillan.

Hemlin, S., Allwood, C. M., & Martin, B. R. (2004). *Creative knowledge environments. The influences on creativity in research and innovation*. Cheltenham, England: Edward Elgar.

Hooghe, L., & Marks, G. (2001). *Multi-level governance and European integration*. Lanham, MD: Rowman & Littlefield Publishers.

Hopwood, B., Mellor, M., & O'Brien, G. (2005). Sustainable development: Mapping different approaches. *Sustainable Development, 13*, 38–52.

in't Veld, R. J. (2010a). *Knowledge democracy. Consequences for science, politics, and media*. Heidelberg, Germany: Springer. https://doi.org/10.1007/978-3-642-11381-9

in't Veld, R. J. (2010b). Towards knowledge democracy. In R. J. in't Veld (Ed.), *Knowledge democracy. Consequences for science, politics, and media* (pp. 1–11). Heidelberg, Germany: Springer. https://doi.org/10.1007/978-3-642-11381-9_1

Inglehart, R. (1990). *Culture shift in advanced industrial society*. Princeton, NJ: Princeton University Press.

IPCC. (2007a). Climate change 2007: The physical science basis – the fourth assessment report of the Intergovernmental Panel on Climate Change by Solomon, S., D. Qin, M. Manning, Z. Chen, M. Marquis, K.B. Averyt, M. Tignor and H.L. Miller (eds.). Cambridge, United Kingdom: Cambridge University Press.

IPCC. (2007b). *Climate change 2007: The physical science basis – errata for the working group - fourth assessment report*. Retrieved from http://www.ipcc.ch/pdf/assessment-report/ar4/wg1/ar4-wg1-errata.pdf

Kaiser, R., & Prange, H. (2004). The reconfiguration of national innovation systems – the example of German biotechnology. *Research Policy, 33*, 395–408.

Kates, R. W., Clark, W. C., Corell, R., Hall, J. M., Jaeger, C. C., Lowe, I., … Faucheux, S. (2001). Environment and development: Sustainability science. *Science, 292*(5517), 641–642.

Kline, S. J., & Rosenberg, N. (1986). An overview of innovation. In R. Landau & N. Rosenburg (Eds.), *The positive sum strategy*. Washington, DC: National Academy Press.

Krausmann, F., Fischer-Kowalski, M., Schandl, H., & Eisenmenger, N. (2008). The global socio-metabolic transition: Past and present metabolic profiles and their future trajectories. *Journal of Industrial Ecology, 12*(5), 637–656.

Kuhlmann, S. (2001). Future governance of innovation policy in Europe – three scenarios. *Research Policy, 30*, 953–976.

Kuhn, T. S. (1962). *The Structure of Scientific Revolutions*. Chicago; The University of Chicogo Press.

Lancet Commission. (2017). The Lancet Commission on pollution and health. *The Lancet*. Retrieved from http://societyforindoorenvironment.net/sites/default/files/pdf/Landriganetal2017TheLancetComissionsHealthAsiaReview.pdf and http://gahp.net/wp-content/uploads/2017/03/PE_InfoLancetSummary.pdf

Laperche, B., & Uzunidis, D. (2018). The knowledge capital of the network firm: Socialization versus business appropriation of scientific work. In G. Bast, E. G. Carayannis, & D. F. J. Campbell (Eds.), *The future of education and labor, ARIS Series: Arts, Research, Innovation and Society*. New York, NY: Springer.

Le Monde diplomatique. (2009). *Atlas der Globalisierung – Sehen und Verstehen, was die Welt bewegt*. Paris, France: Le Monde diplomatique/taz.

Leydesdorff, L. (2012). The Triple Helix, Quadruple Helix, …, and an N-tuple of Helices: Explanatory models for analyzing the knowledge-based economy? *Journal of the Knowledge Economy, 3*(1), 25–35. https://doi.org/10.1007/s13132-011-0049-4

Lundvall, B.-Å. (Ed.). (1992). *National systems of innovation. Towards a theory of innovation and interactive learning*. London, UK: Pinter Publishers.

Malerba, F. (1999). *Sectoral systems of innovation and production*. DRUID Conference on National Innovation Systems, Industrial Dynamics and Innovation Policy (Rebild, June 9–12, 1999). Retrieved from http://www.druid.dk/uploads/tx_picturedb/ds1999-69.pdf

Malerba, F. (2002). Sectoral systems of innovation and production. *Research Policy, 31*(2), 247–264.

Malerba, F. (2004). *Sectoral systems of innovation: Concepts, issues and analyses of six major sectors in Europe*. Cambridge, England: Cambridge University Press.

Meglic, J., Kern, T., Urh, B., Balkovec, J., & Roblek, M. (2009). Influence of polyvalence professionals on product development process efficiency. *Strojarstvo, 51*(2), 105–121.

Merz, M., & Sormani, P. (rédacteurs). (2016). *The local configuration of new research fields. On regional and national diversity*. Cham, Switzerland: Springer.

Milbergs, E. (2004). *Measuring innovation for national prosperity*. Washington, DC: Center for Accelerating Innovation. Retrieved from http://www.innovationecosystems.com/images/Innovation_Framework_Report_3.00_January_2004_Update.pdf

Milbergs, E. (2005a). *Innovation ecosystems and prosperity*. Washington, DC: Center for Accelerating Innovation. Retrieved from http://www.innovationecosystems.com/images/GW_Presentation_March_29_2005_Final_Nnotes_1.1.pdf

Milbergs, E. (2005b). *Setting the stage for innovation. Accelerating innovation 2005 conference*. Washington, DC: Center for Accelerating Innovation. Retrieved from http://www.innovationecosystems.com/images/Accelerating_Innvation_October_19_version_3.0.pdf

Mitterlehner, B. (2014). Cyber-democracy and cybercrime: Two sides of the same coin. In E. G. Carayannis, D. F. J. Campbell, & M. P. Efthymiopoulos (Eds.), *Cyber-development, cyber-democracy and cyber-defense. Challenges, opportunities and implications for theory, policy and practice* (pp. 207–230). New York, NY: Springer.

Miyata, Y. (2003). An analysis of research and innovative activities of universities in the United States. In L. V. Shavinina (Ed.), *The international handbook on innovation* (pp. 715–738). Amsterdam, Netherlands: Pergamon.

Monfreda, C., Wackernagel, M., & Deumling, D. (2004). Establishing national natural capital accounts based on detailed ecological footprint and biological capacity assessments. *Land Use Policy, 21*(3), 231–246.

Narin, F., Hamilton, K. S., & Olivastro, D. (1997). The increasing linkage between U.S. technology and public science. *Research Policy, 26*, 317–330.

National Science Board. (2008). *Science and engineering indicators 2008* (Vols. 1 and 2). Arlington, VA: National Science Foundation. Retrieved from http://www.nsf.gov/statistics/seind08

Nelson, R. R. (1990). Capitalism as an engine of progress. *Research Policy, 19*, 193–214.

Nelson, R. R. (Ed.). (1993). *National innovation systems. A comparative analysis*. Oxford, England: Oxford University Press.

Nowotny, H., Scott, P., & Gibbons, M. (2001). *Re-thinking science. Knowledge and the public in an age of uncertainty*. Cambridge, England: Polity Press.

Nowotny, H., Scott, P., & Gibbons, M. (2003). Mode 2 revisited: The new production of knowledge. *Minerva, 41*, 179–194.

Nowotny, H., Scott, P., & Gibbons, M. (2006). Re-thinking science: Mode 2 in societal context. In E. G. Carayannis & D. F. J. Campbell (Eds.), *Knowledge creation, diffusion, and use in innovation networks and knowledge clusters. A comparative systems approach across the United States, Europe and Asia* (pp. 39–51). Westport, CT: Praeger.

O'Donnell, G. (2004). Human development, human rights, and democracy. In G. O'Donnell, J. V. Cullell, & O. M. Iazzetta (Eds.), *The quality of democracy. Theory and applications* (pp. 9–92). Notre Dame, IN: University of Notre Dame Press.

Obama, B. (2017). The irreversible momentum of clean energy. *Science*. https://doi.org/10.1126/science.aam6284. Retrieved from http://science.sciencemag.org/content/early/2017/01/06/science.aam6284.full

OECD. (1994). *The measurement of scientific and technological activities. Proposed standard practice for surveys of research and experimental development. Frascati manual 1993*. Paris, France: Author.

OECD. (2002). *The measurement of scientific and technological activities. Proposed standard practice for surveys on research and experimental development. Frascati manual 2002*. Paris, France: Author.

Palmberg, C., & Luukkonen, T. (2006). The different dynamics of the biotechnology and ICT sectors in Finland. In E. G. Carayannis & D. F. J. Campbell (Eds.), *Knowledge creation, diffusion and use in innovation networks and knowledge clusters* (pp. 158–182). Westport, CT: Praeger.

Park, H. W. (2014). Transition from the Triple Helix to N-Tuple Helices? An interview with Elias G. Carayannis and David F. J. Campbell. *Scientometrics 99*(1), 203–207. Retrieved from http://link.springer.com/article/10.1007%2Fs11192-013-1124-3 and http://download.springer.com/static/pdf/907/art%253A10.1007%252Fs11192-013-1124-3.pdf?auth66=1397308723_4cb0003877af5305d5dc202280b9cd6d&ext=.pdf

Pfeffer, T. (2012). *Virtualization of universities. Digital media and the organization of higher education institutions*. New York, NY: Springer. Retrieved from http://www.springer.com/business+%26+management/media+management/book/978-1-4614-2064-4

Plasser, F., & Plasser, G. (2002). *Global political campaigning. A worldwide analysis of campaign professionals and their practices*. Westport, CT: Praeger.

Polanyi, M. (1962). The republic of science: Its political and economic theory. *Minerva 1*, 54–74. Retrieved from http://sciencepolicy.colorado.edu/students/envs_5100/polanyi_1967.pdf and http://fiesta.bren.ucsb.edu/~gsd/595e/docs/41.%20Polanyi_Republic_of_Science.pdf

Rapport, D. J. (2007). Sustainability science: An ecohealth perspective. *Sustainability Science, 2*, 77–84.

Saint-Paul, G., & Verdier, T. (1993). Education, democracy and growth. *Journal of Development Economies, 42*, 399–407.

Schafer, K. E. (2008). Have we succeeded? Political leadership lessons from the United States. In A. Zimmer & R. M. Jankowitsch (Eds.), *Political leadership. Annäherungen aus Wissenschaft und Praxis* (pp. 263–278). Berlin, Germany: Polisphere.

Schumpeter, J. A. (1976/first published in 1942). Capitalism, socialism and democracy. New York, NY: Harper Perennial.

Schumpeter, J. A. (1934). *The theory of economic development*. Cambridge, Massachusetts: Harvard University Press.

Shavinina, L. V. (Ed.). (2003). *The international handbook on innovation*. Amsterdam, Netherlands: Pergamon Elsevier Science.

Steffen, W., Rockström, J., Richardson, K., Lenton, T. M., Folke, C., Liverman, D., Summerhayes, C. P., Barnosky, A. D., Cornell, S. E., Crucifix, M., Donges, J. F., Fetzer, I., Lade, S. J.,

Scheffer, M., Winkelmann, R., & Schellnhuber, H. J. (2018). Trajectories of the earth system in the anthropocene. *Proceedings of the National Academy of Sciences of the United States of America 115*, 8252. https://doi.org/10.1073/pnas.1810141115. Retrieved from http://www.pnas.org/content/early/2018/08/07/1810141115/tab-article-info and http://www.pnas.org/content/pnas/early/2018/08/07/1810141115.full.pdf\

Tassey, G. (2001). R&D policy models and data needs. In M. P. Feldman & A. N. Link (Eds.), *Innovation policy in the knowledge-based economy* (pp. 37–71). Boston, MA: Kluwer Academic Publishers.

U.S. Census Bureau. (2008). *Voting and registration in the election of November 2008*. Washington, DC: Author. Retrieved from http://www.census.gov/hhes/www/socdemo/voting/publications/p20/2008/Table%2005-1.xls

Umpleby, S. A. (2005). What I learned from Heinz von Foerster about the construction of science. *Kybernetes, 34*(1-2), 278–294.

UNDP. (2007). Human development report 2007/2008. *Fighting climate change: Human solidarity in a divided world*. New York, NY: Author. Retrieved from http://hdr.undp.org/en/reports/global/hdr2007-2008

UNDP. (2009). Human development report 2009. *Overcoming barriers: Human mobility and development*. New York, NY: Author. Retrieved from http://hdr.undp.org/en/reports/global/hdr2009

UNEP. (2008). *Green jobs: Towards decent work in sustainable, low-carbon world*. Washington, DC: Author. Retrieved from http://www.unep.org/labour_environment/PDFs/Greenjobs/UNEP-Green-Jobs-Report.pdf

United Nations. (1987a). *Report of the World Commission on Environment and Development (42/187)*. New York, NY: Author. Retrieved from http://www.un-documents.net/a42r187.htm

United Nations. (1987b). *Report of the World Commission on Environment and Development: "Our common future"*. New York, NY: Author. Retrieved from http://worldinbalance.net/pdf/1987-brundtland.pdf

Vadrot, A. (2008). Der Wille zur Wahrheit als der Wille zur Macht. *Die theoretischen Implikationen der Schnittstelle zwischen Politik und Wissenschaft in der Biodiversitätspolitik*. (Magisterarbeit., University of Vienna, Vienna, Austria).

Von Hippel, E. (1988). *The sources of innovation*. Oxford, England: Oxford University Press. Retrieved from http://web.mit.edu/evhippel/www/sources.htm

Von Hippel, E. (2005). *Democratizing innovation*. Cambridge, MA: MIT Press. Retrieved from http://web.mit.edu/evhippel/www/democ1.htm

Winiwarter, V., & Knoll, M. (2007). *Umweltgeschichte*. Köln, Germany: Böhlau.

World Meteorological Organization. (2017). The sate of greenhouse gases in the atmosphere based on global observations through 2016. *WMO Greenhouse Gas Bulletin* 13 (October 30, 2017). Retrieved from https://ane4bf-datap1.s3-eu-west-1.amazonaws.com/wmocms/s3fs-public/ckeditor/files/GHG_Bulletin_13_EN_final_1_1.pdf

Index

E. G. Carayannis, D. F. J. Campbell, *Smart Quintuple Helix Innovation Systems*, SpringerBriefs in Business, https://doi.org/10.1007/978-3-030-01517-6

The manufacturer's authorised representative in the EU is Springer Nature Customer Service Centre GmbH, Europaplatz 3, 69115 Heidelberg, Germany. If you have any concerns regarding our products, please contact ProductSafety@springernature.com

Printed and bound by CPI Group (UK) Ltd, Croydon, CR0 4YY

30/09/2026

02225606-0005